KAMUT®

An Ancient Food

For A Healthy Future

A grain leading the green revolution
for easy access to
vitamins, minerals, enzymes,
and hormone precursors

Betty Kamen, Ph.D.

N_E

Nutrition Encounter, Novato, California

All of the facts in this book have been very carefully researched and have been drawn from the scientific litera-ture. In no way, however, are any of the suggestions meant to take the place of advice given by physicians. Please consult a medical or health professional should the need for one be indicated.

Kamut is a registered trademark of Kamut International, Ltd.

Nutrition Encounter, 1995
Box 5847
Novato, CA 94948

Printed in the United States of America
First Printing: 1995

ISBN 0-944501-09-5

Dedicated to

David Sandoval

Who looked at the past,

had a vision for the future,

and made that vision a reality.

OTHER BOOKS BY BETTY KAMEN, PH.D.

Total Nutrition During Pregnancy:
How To Be Sure You and Your Baby
Are Eating the Right Stuff

Total Nutrition for Breast-Feeding Mothers

Kids Are What They Eat:
What Every Parent Needs to Know About Nutrition

In Pursuit of Youth: Everyday Nutrition

Osteoporosis: What It Is, How to Prevent It, How to Stop It

Nutrition In Nursing: The New Approach
A Handbook of Nursing Science

Sesame: The Superfood Seed
How It Can Add Vitality To Your Life

Siberian Ginseng:
Up-To-Date Research on the Fabled Tonic Herb

Germanium: A New Approach to Immunity

Startling New Facts About Osteoporosis:
Why Calcium Alone Does Not Prevent Bone Disease

The Chromium Connection: A Lesson in Nutrition

New Facts About Fiber:
How Fiber Supplements Can Enhance Your Health

Everything You Always Wanted to Know About Potassium
But Were Too Tired to Ask

Hormone Replacement Therapy: Yes or No?
How to Make an Informed Decision

Betty Kamen is an award-winning photojournalist with graduate degrees in psychology and nutrition education. She is an internationally-known lecturer, radio/TV host, and author of many major books, hundreds of articles, and several tapes on various aspects of health and nutrition. For many years she hosted *Nutrition 57* on WMCA in New York; *Nutrition Dialogue,* SPN TV Cable Network; followed by *Nutrition Watch* on KNBR in San Francisco.

CONTENTS

1

THE PYRAMID STORY

Reaching Back to Antiquity

Portugal — 1949

A young United States airman, far from his home in the American heartland, wandered into a shop and engaged the merchant in conversation about farming, weather, and home. He could think of no better activity for relieving his sense of loneliness than to find an interesting gift to send back to the folks in Montana. The merchant dealt with this sort of customer before and easily rose to the occasion.

"This was left here by a scientist before the war," explained the merchant in halting English as he reached for a curious stone box on a high shelf toward the back of the store. "He was fleeing from Egypt — an archaeologist, I think he was. He was afraid he would be in trouble with the authorities if he had an ancient artifact in his possession."

The serviceman was only mildly interested in the story, concluding that the merchant was no doubt full of similar anecdotes — perhaps altered to fit each customer's need. But, not to be rude, he listened.

"So the scientist left the box with me," continued the merchant. "I helped him with his arrangements to get home, and he promised to return to pick up the box, which he asked me to guard carefully. He said the contents of the box came from a tomb in Egypt and that it was 5,000 years old." Continuing in a half whisper and with much drama, he added, "That was almost ten years ago, and he has not come back yet!"

The merchant looked around furtively, as though to be certain there were no eavesdroppers — perhaps even to rule out the possibility of a nearby thief or spy.

> Satisfied that conditions were secure, or perhaps that he now had the airman's attention, he slowly pried open the lid of the strange little box.

The airman, until now just humoring the merchant, suddenly became intensely interested. It wasn't the story that captured his curiosity, but rather the contents of the box. No old coins, no jewelry, no gold — but seeds! Thirty-six kernels of wheat! Strange, however, because these weren't like the wheat kernels with which he was familiar — the ones that had been so much a part of his boyhood years on the family farm in Montana. No — these were more than twice as big and had an unusual humpback shape.

A small paper label on the inside lid of the box read, "Excavated from a tomb in the pyramid Saqqar, near Dahshur." Although he could hardly understand his own reaction, the airman could not conceal an overwhelming and profound sense of excitement.

It's not hard to imagine what happened next. The merchant asked an exorbitant price. The airman countered with a far lower bid, rejected with disdain by the merchant. As the airman's offer escalated, the merchant's price remained fixed. The airman upped his bid again, his demeanor more anxious with each accession. The merchant, lowering his request only slightly from time to time, did so in a manner denoting total control of the negotiations. After hours, followed by days of bickering — some say there was even a wager involved — the airman was in possession of the strange little box containing the extraordinary kernels.

He sent the seeds home to Montana with a letter explaining their ancient origin. When his father received the unusual package, he planted the seeds — curious to see what, if anything, would happen. To his surprise, thirty-two of the seeds germinated!

Six seasons later the farmer had 1,500 bushels of this "new" strain of wheat filling his granary.

Montana — 1955
The farmer was a big hit at the county fair with "King Tut's Wheat," as this grain had come to be known. In addition to the uncommonly large kernels, King Tut's Wheat had a rich buttery taste and made a fine pasta. But there was no apparent commercial market for the product. After the fair, the remaining stock was used for animal feed.

Montana — 1977
The grain was forgotten for two decades, until agricultural scientist Dr. Bob Quinn decided to follow up on something he remembered seeing at the county fair in his youth. Working with his father, Mack Quinn, he managed to locate a jar of "King Tut's Wheat." They brought the grain to their own wheat and cattle ranch near Big Sandy, Montana. It was the Quinns who dubbed their find "kamut" (pronounced ka-MOOT), after an ancient Egyptian word for wheat.

Washington, D.C. — 1990
The Quinns were granted Variety Protection Certificates from the United States Department of Agriculture, which identified and recognized the grain as KQ-77, a specific and protected variety. Kamut is now a registered trademark of Kamut International, Ltd.

Can we really trace the lineage of today's commercial kamut to seeds that germinated after being sequestered in a pyramid for 5,000 years? It's a great story, perhaps even worthy of a grade B science fiction movie. But I've dealt with enough small shop owners in remote parts of the world (from primitive Balikpapan to bustling Kuala Lumpur to the offbeat left bank of Paris) to know that the chances of the story being true are, at best, suspect. We do know, however, that thirty-six kernels were sent to Montana from somewhere in Europe, and that the grain now being sold as kamut is derived from these kernels. And there have been at least a few documented cases of grains from ancient tombs germinating successfully, some even recently.

We also know that similar grains have been under more-or-less continuous cultivation in parts of the Middle East since the beginning of civilization. So it isn't really necessary to look to the story of an ancient Egyptian tomb for an explanation of how a handful of unusual seeds found their way to Montana in 1949! Nevertheless, the "time capsule" story conveys the concept of this grain's relationship to other grain and wheat products.

The story is accurate in spirit, if not precise in fact.

2

SOME BOTANY

Why Kamut is Better

Just how *does* kamut relate to other grains and to other
kinds of wheat? Along with rice, corn, barley, rye, oats, and
millet, the wheat genus (Triticum) is a member of the grass
family (Gramineae). Different types of wheat are further
categorized depending on how many chromosomes they
contain in each cell. Chromosomes are microscopic rod-
shaped bodies that contain the genes that carry all the
detailed genetic information of the organism. Each living
cell in the organism has the same characteristic number of
chromosomes.

 Chromosomes give cells
the direction they need for
reproducing themselves —
whether in humans or
plants.

The most ancient type of wheat, the *Diploid* group, has two sets of seven chromosomes arranged in pairs for a total of fourteen. (The *di* prefix refers to *two*.) This wheat is found wild but seldom cultivated. Einkorn (*Triticum monococcum*) is one of this group. It's not a name familiar to me, and, unless you are a wheat farmer, probably not to you either.

Tetraploid wheats, with four seven-chromosome sets, totalling twenty-eight, (*tetra* meaning *four*), also grow wild but have been cultivated extensively for thousands of years and include emmer, durum, Persian, and poulard wheat. These names are somewhat familiar to those of us who have been interested in alternatives to common wheat (for reasons explained later).

Kamut is one of the tetraploids, technically called *Triticum durum*.

The *hexaploid* group (*hexa* for *six*), with six sets of seven chromosomes for a total of forty-two, has a history almost as mysterious as the thirty-six kernels of kamut. It appeared about 8,000 years ago (not a very long time on an evolutionary scale measured in millions of years), either as a genetic accident or as the result of deliberate crossing of two other types of grain. Either way, the cultivation of wheat with those extra seven pairs of chromosomes is thought to have played a major role in the viability of ancient civilizations.[1] Modern common wheat, *Triticum vulgare*, is a hexaploid, as is club wheat, spelt, shot, macha, and valvilovii.

Perhaps the most important fact about kamut is that it hasn't been subjected to millennia of "improvement" by the application of agricultural technology.

Just as spelt is an ancient ancestor of today's common bread wheat and has remained relatively unaltered, kamut is an ancestor of modern durum wheat. Common bread wheat and modern durum wheat remind me of commercial tomatoes. You know, those bright red things you see in the supermarket whose genes have been manipulated until they are tough enough to survive a twelve-mile-per-hour impact against a brick wall. Just don't judge them for taste because they have none! Spelt and kamut, on the other hand, are reminiscent of the tomatoes you grow in your garden — lower in yield and lacking in rugged handling and shipping properties, but downright delicious.

Even the distinction between spelt and kamut, however, is important. Spelt is one of the hexaploid group with forty-two chromosomes. Kamut, on the other hand, retains the "natural" twenty-eight chromosomes of the tetraploid group.

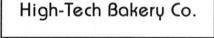

High-Tech Bakery Co.

No, Sir, we're not worried about the farmer's strike. We haven't used any real grains in our baked goods for some time.

TABLE 1

Comparison of Kamut Grain, Common Wheat (US averages), and Shredded Wheat

Nutrient	Kamut Grain	Common Wheat	Shredded Wheat
Water	9.8%	11.5%	6.7%
Protein	17.3%	12.3%	9.9%
Total fat	2.6%	1.9%	2.0%
Carbohydrates	68.2%	72.7%	75.0%
Crude fiber	1.8%	2.1%	na
Ash	1.8%	1.6%	na
Calories/100 g	359	335	364
Minerals (mg/100g)			
Iron	4.2	3.9	3.5
Magnesium	153.0	117.0	142.0
Phosphorus	411.0	396.0	389.0
Potassium	446.0	400.0	346.0
Sodium	3.8	2.0	3.0
Zinc	4.3	3.2	na
Copper	0.4	0.4	0.18
Manganese	3.2	3.8	na
Vitamins (mg/100g)			
Thiamin	0.45	0.42	0.21
Riboflavin	0.12	0.11	0.11
Niacin	5.54	5.31	4.39
Pantothenic acid	0.32	0.91	0.71
Pyridoxine (B6)	0.08	0.35	0.25
Folacin	0.03	0.04	0.07
Vitamin E	1.70	1.20	0

Lipids (g/100g)

Saturated (16:0 Palmitic acid)	0.550	0.303	na
Monounsaturated (18:1 Oleic acid)	0.400	0.225	na
Polyunsaturated (18:2 Linoleic acid)	1.580	0.733	na
Polyunsaturated (18:3 Linoleic acid))	0.125	0.035	na
Cholesterol	0	0	na

Amino Acids (g/100g)

Tryptophan	0.117	0.194	na
Threonine	0.540	0.403	na
Isoleucine	0.060	0.520	na
Leucine	1.230	0.964	na
Lysine	0.440	0.361	na
Methionine	0.250	0.222	na
Cystine	0.580	0.348	na
Phenylalanine	0.850	0.675	na
Tyrosine	0.430	0.404	na
Valine	0.800	0.624	na
Arginine	0.860	0.610	na
Histidine	0.430	0.312	na
Alanine	0.630	0.491	na
Aspartic acid	0.980	0.700	na
Glutamic acid	5.970	4.680	na
Glycine	0.065	0.560	na
Proline	1.440	1.500	na
Serine	0.930	0.662	na

* NA = Values Not Available

Source: *Agriculture Handbook No. 8-20, Cereal Grains and Pasta*, 1988, pp 2071-2076; Medallion Laboratory Analytical Report No. 88011589, prepared for Montana Flour & Grains on Kamut, October 24, 1988.

Table 1 on pages 8 and 9 compares the nutrient composition of kamut grain with common wheat. Many of the lower values of common wheat are a result of selective breeding for drought resistance. This is great for the farmer but does nothing for nutritional value. Kamut has 80 percent more fat — much of it from the valuable linoleic acid — and a more desirable mix of amino acids. Common wheat, on the other hand, has 66 percent more tryptophan, despite the reduced overall protein percentage. But farmers don't care about amino acid ratios, so what we get is a near-random result of careful selection for other attributes (the profit motive at work) with very little regard for human nutrition — not exactly in your best health interest.

The items listed in Table 1 are only the easily measured effects of selective breeding and hybridization. Surely there are variations in many other physiological factors. The implications of these changes on human health and nutrition are still very poorly understood. Wheat allergies are widespread, yet often subtle and undiagnosed. Although both types of grain provide dietary gluten — the substance usually blamed for wheat allergy — there are significant differences in how the two types of grain are metabolized.

Researchers are beginning to pay attention to these differences. In a recent paper published in Italy, it was noted that tetraploid durum wheat does not have the toxic effect on certain types of small intestinal lesions attributed to hexaploid bread wheat.[2]

Kamut has an exquisite protein advantage. Very few plant foods contain all eight essential amino acids.

The term *essential* in nutrition language means it must be on your plate because your body can't produce it.

IMPORTANT!

Look what happens to a cereal grain with processing. (See the Shredded Wheat column in Table 1.) No self-respecting bug attacks the boxes of cold cereal sitting on the pantry shelf. Why should they? It's dead food, lacking in enzymes, missing other "life" properties. But among the embalmed array of cold cereals that we mistakenly refer to as "food," shredded wheat is a shining star — the best among the worst. Note, however, that it has less protein, iron, potassium, magnesium, and β vitamins than the wheat from which it was spawned.

We have come to accept all kinds of food as "normal," even though they barely resemble their original architecture.

Throughout the history of western civilization, wheat has been selected for high yield and resistance to harsh weather and pests, plus ease of production and processing. In the past century the advent of modern agricultural techniques has accelerated the selection process. In the past decade, genetic engineering has sent the technology into orbit, figuratively and literally: Biotech experiments have even been performed on the space shuttle — again, with convenience and high profits as the goals!

Although we seek variety on the dinner table, the original sources of most of our foods are the same. More and more plants are falling into disuse. The United Nations Food and Agriculture Organization, which monitors the world's plant and animal life, calls this "a catastrophe in the making." Despite up to 50,000 edible plants in the world today, rice, corn, and wheat account for 60 percent of the proteins and calories derived from plants. One-third of American prairies are planted under only one variety of wheat today. The problem is that monoculture subjects us to outbreaks of disease and pests.[3]

Surely you've read articles calling attention to the dangers of narrowing the diversity of our growing fields.

Kamut has escaped this selective breeding. Through scarcity and obscurity (even if it *wasn't* preserved for 5,000 years in an Egyptian tomb), we have a wheat product that appears to be a lot closer to the kinds of foods our bodies were genetically designed to thrive on.

Is there a health benefit from substituting kamut for other forms of wheat? Yes, especially for those who have some degree of allergy to common bread wheat, as many people do. In fact, the vast majority of those with wheat allergies have no idea that this is the source of many of their chronic health problems. (We'll discuss this in detail later.) The extraordinary health advantage, however, is not only the substitution of one grain for another, *but the use of a nutritionally potent form of the kamut plant.*

The dark green leafy part of the young kamut plant is one of the most healthful foods on the planet, rivaling any vegetable on your table for nutritive value. By the time the plant matures to the "amber waves of grain" stage, most of the best nutrients have been utilized in the plant's seed production — the seed has to be fully equipped to propagate the next generation. The grains of the grass family are far more valuable in their leafy green stage of growth as a *vegetable* than they are in the seed stage as a source of *flour* for baked goods and pasta.

The development and survival of food products are regulated by mechanisms that may bear little resemblance to the world in which we live. The path of their evolution can be affected by major events, and sometimes by events that may be unnoticeable.[4] But I want to call everyone's attention to a remarkable set of circumstances — the specific reasons for kamut's superiority.

You already know that kamut seeds found their way from somewhere abroad to a farm in Montana. Today, growth starts on top of a mountain, 5,500 feet high, on an ancient volcanic lake bed. The soil is mineral-rich, pure-watered by the subterranean lake bed — never contaminated; benefiting from the snow's runoff in the summer; always fresh and clean, providing a steady renewal of nature's true force — a rarity on our troubled earth.

The cold nights of the high atmosphere
add to the natural process, creating a
plant with a full complement of high-qual-
ity nutrients.

The seed selection is nonhybrid, 100 percent organic, pes-
ticide-free — the only one of its kind. The soil has been
certified as organic — no chemical fertilizers to destroy
nutrients here! Kamut thrives in this totally natural world —
a world of long ago surfacing again as we approach the
twenty-first century.

The elevation is above the atmosphere that ordinarily filters
the sun — so the leaves of the kamut plant, as they face
upward, are privy to greater concentrations of sunlight than
their peers down below. Because of this greater exposure,
the production of chlorophyll is increased (a process explained later). The higher altitude also diminishes oxygen.
To compensate, the plants work harder and produce more
protein, as though doing their nautilus workouts.

Kamut is an annual plant. No hybridization to force more frequent breeding has
been applied. Kamut lives and grows from
sprouts to leaves to seeds in one year. It
is harvested seasonally, following the in-
scrutable wisdom of nature.

This wonderful environment imparts its goodness to the
plant, just as some greenhouse-grown plants are disadvan-
taged because of their artificial atmosphere — hydroponics,
forced unnatural lighting, manually controlled temperature,
and so on — all taking their toll on nutrient values.

Processing takes place in close proximity to the growing location — right in the field. The plants are moved rapidly from field to pristine processing room, where they are washed to remove surface particles and then sent through a gentle extraction operation.

About 99 percent of the fiber is removed, and the remaining juice is instantly placed in a special vacuum dehydrator. The finished product contains levels of chlorophyll and protein that are significantly superior to any other dehydrated cereal juice powder. What remains is 100 percent organic dried juice powder — again, the only one of its kind!

Impressed? So was I! In fact, I've joined the Kamut Lovers of the World and it's *been a soothing, strengthening, cleansing, and healing trip* — the very reasons this book has been written!

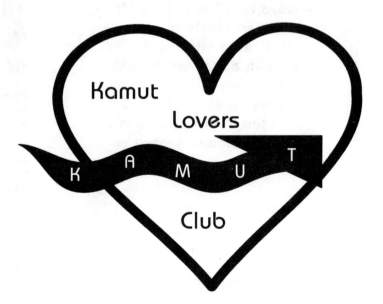

IMPORTANT!

Comparison Summary of Kamut and Common Wheat

Kamut has:

- ~ 29% more protein
- ~ 27% more lipids
- ~ 23% more magnesium
- ~ 25% more zinc
- ~ more riboflavin, thiamin, and niacin
- ~ much more vitamin E
- ~ sixteen amino acids with higher values
- ~ higher values in eight of the 9 minerals found in common wheat

3

SOME FARMING

And How It Affects Your Well-being

From Seed to Sprout

From the time a seed germinates, it has one purpose in life: *to produce more seeds!*

The process begins with the pushing of a short blade of grass above the soil to capture solar energy.

If the grain is a winter variety, this happens in the fall, but the plant remains mostly dormant until spring. It germinates and survives the snow cover, but nothing much more occurs. (This is very different from what happens in a hothouse, where the plant is confused into thinking it's *not* winter and continues to grow.)

The leaves grow quickly, using sunlight to make sucrose from carbon dioxide and water. This is the basis for countless subsequent chemical processes that manufacture fats, amino acids, proteins, complex carbohydrates, and even enzymes that control these processes. The machinations of this solar-powered factory in the leaves of the young grass plant are almost beyond comprehension in their complexity.

Results depend heavily on environmental factors. Without key minerals from the soil and the right changes in climate, the sprouted seed misses the enzymes needed to enter the next growth phase.

Winter grasses grown in a greenhouse, although appearing to flourish, lack enzymes, proteins, and amino acids necessary to develop into seed-bearing mature plants. The leaves will manufacture sugar through photosynthesis but cannot go into full production of the complicated organic molecules required for reproduction. Grains sprouted without access to minerals from soil lack the total gamut of raw materials necessary for producing seed.

This is not to say that the sprouts you grow in your kitchen (without benefit of soil) are not a valuable food. Because of their extreme freshness, I consider sprouts a vital part of my everyday diet. Sprouts are virtually the only food living and growing right up to the time they enter my mouth. They're also completely free of pesticides, additives, and other forms of food pollution — both intentional and incidental. So I'm not about to dismiss the importance of homegrown sprouts in the modern diet.

Children particularly enjoy home sprouting because they can almost *see* the unborn, sleeping little stores of energy as the dormant seeds urge forward to life, slowly but surely.

IMPORTANT!

Advantages of Sprouted Seeds

~ Enzymes are activated.

~ Proteins convert to free amino acids.

~ Starches change to simple plant sugars.

~ Minerals combine to increase assimilation.

~ Vitamin content increases from 3 to 12 or more times.

~ Chlorophyll and carotene content increase dramatically when exposed to sunlight.

TABLE 2

Comparison of Kamut Grain, Dehydrated Cereal Grass & Dehydrated Kamut Juice with Alfalfa Juice

Nutrient	Kamut Grain	Dehydrated Cereal Grass	Dehydrated Kamut & Alfalfa Juice
Water	9.8%	na	1.16%
Protein	17.3%	23.0%	24.94%
Total Fat	2.6%	na	na
Carbohydrate	68.2%	37.1%	na
Crude fiber	1.8%	37.5%	na
Ash	1.82%	na	na
Chlorophyll	na	0.54%	3.10%
Calories/100 g	359	286	na

MINERALS (mg/100g)

Nutrient	Kamut Grain	Dehydrated Cereal Grass	Dehydrated Kamut & Alfalfa Juice
Calcium	31.0	514	698
Iron	4.2	57	45
Magnesium	153.0	103	610
Phosphorus	411.0	514	844
Potassium	446.0	3,200	3,660
Sodium	3.8	29	592
Zinc	4.3	0.5	na
Copper	0.4	0.6	1.2
Manganese	3.2	10	15
Silicon	na	na	784
Boron	na	na	17

VITAMINS (mg/100g)

Thiamin (B1)	0.45	0.29	21.4
Riboflavin (B2)	0.12	2.00	6.4
Niacin	5.54	7.50	na
Pantothenic Acid	0.32	2.40	386.0
Pyridoxine (B6)	0.08	1.30	98.0
Folacin	0.03	1.10	na
Vitamin E	1.70	0.03	na
Choline			5,360
Beta-carotene (IU/100g)		23,000	26,000

AMINO ACIDS (g/100g)

Tryptophan	0.117	0.486
Threonine	0.540	na
Isoleucine	0.0600	1.178
Leucine	1.23	2.244
Lysine 0.440	1.516	na
Methionine	0.250	1.7476
Cystine	0.580	0.3634
Phenylalanine	0.850	1.4300
Tyrosine	0.430	0.8950
Valine	0.800	1.4620
Arginine	0.860	1.4860
Histidine	0.430	0.5088
Alanine	0.630	1.8160
Aspartic acid	0.980	2.6640
Glutamic acid	5.97	3.0040
Glycine	0.065	1.5200
Proline	1.44	1.5100
Serine	0.930	1.2100

Source: Seibold, RL. *Cereal Grass, What's In It For You!* (Lawrence, Kansas: Wilderness Community Education Foundation, 1990).

A truly healthy plant is able to reproduce itself while many home-sprouted or greenhouse-grown grains cannot. It's like the difference between the eggs from the health store, which would have become chicks if fertilized, and the commercial eggs from the supermarket, which are far from fertile because of nutrient deficiencies. We can even taste the difference! Another example occurs at Thanksgiving. Virtually all mass-farmed turkeys have lost the ability to procreate. Without artificial insemination, they would vanish after a single generation.

Human fertility is analogous. Fifteen percent of Americans cannot reproduce, even though these couples appear to be healthy. In a four-year period, the average sperm count decreased by 30 percent!

The nutrients in the leaves of the young grass plant reach maximum concentration just before a phase in the plant's development called the *jointing* stage. This is when the stem begins to develop and chlorophyll, protein, and vitamin content decline while fiber content increases.[5] The reservoir of organic chemicals and special nutrients in the leaves are channelled into the seed kernels, which is why these chemicals were manufactured in the first place.

Green Versus Grain
So you can see why the young kamut grass plant has a higher nutrient value than the grain. While the grain is superior to other grains in many ways, it's the *green* that really makes the difference. Look at the amazing accumulation of vitamin A, potassium, and calcium found in kamut grass but absent in the grain. (See Table 2 on pages 20 and 21.) Combine these nutrients with chlorophyll and other cofactors and the greens are worth a king's ransom.

Quietly, I play a game when I attend the major health food industry conventions. As I meet new people, before I know anything about them, I try to conjecture about what their diets are like. I've found that, of all the many dietary variations, there's only one that I can usually guess correctly: *the consumption of lots of green foods!* If they do eat greens, I note a kind of vibrancy, a look of youth and energy that is reflected in skin tone and eyes. This is particularly evident in those who partake of green juices and related concentrated products. What I see is nothing that could be measured scientifically, but I'm convinced that it's there.

Dehydrated foods evoke memories for me. I can remember seeing strings suspended across my grandmother's kitchen ceiling every fall. As a child, I stared wide-eyed at the apple pieces held hostage by clothespins. Dancing at the slightest provocation (frequently encouraged by older male cousins), the apple slices sent an aroma through the house, making our mouths water.

Although I was told the finished bounty was not to be eaten then, but to be stored until the dead of winter, I managed to sneak a piece here and there. I still recall my sense of wonder as to why the stolen little pieces of dried apple were so much more pungent than fresh apples "in season." Years later, food composition handbooks satisfied my curiosity, revealing the higher values of the dried apple's concentrated nutrients, and consequently more piquant flavor.

Just check the vitamin and mineral status of any food and you can't help but notice that dried foods in general rank far and above the "regulars" in most categories — with the possible exception of vitamins C and A.

Now look again at Table 2 on pages 20 and 21, and you can't help but notice the amazing difference. Most of the water content is, of course, removed in the drying process. So on a portion-for-portion basis, the dried food has a significantly increased density of many nutrients, particularly the more stable minerals.

One of the oldest methods of food preservation is now in step with late twentieth century lifestyle —*it's convenient!* You don't have to hang strings from the ceiling.

Drying retains more nutritional value than toasting, roasting, steaming, baking or frying, mainly because the temperatures used are so low.

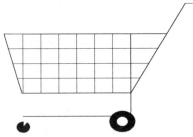

The advent of the supermarket was the worst thing that ever happened to a vegetable. Dehydrated kamut juice helps to supersede that tragedy.

What About Wheat Grass Grown Indoors?

Indoor wheat grass never reaches the jointing stage and never produces a grain kernel. No wonder! It's deprived of the minerals and the environmental triggers of temperature and humidity found in soil: *it's sterile.* (No, it's not devoid of nutrients, but just isn't healthy enough to reproduce, as described earlier.) Wheat grass grown indoors has been summarized in a publication distributed by the Wilderness Community Education Foundation as follows:

> This wheat grass grows quite differently from the wheat grass planted in the ground to produce grain. Although it is a lovely and useful green plant, it does not develop deep roots, absorb soil nutrients, or pass through the growth stages necessary to produce the nutritionally potent wheat grass[6]....

Not unlike kitchen sprouts, I consider tray-grown wheat grass and the juice made from the leaves to be a valuable therapeutic food. But don't confuse it with the real thing!

Green Vegetables and the Garbage Study

A few years ago, a team of researchers at a university in Chicago decided to treat the municipal dump as if it were an archaeological dig. The object was to scientifically evaluate food waste to see what inferences could be made about the eating habits of the society that produced the garbage.

The study makes fascinating reading for a number of reasons, but I mention it because of one important finding: A survey, with participants trying to accurately describe their own diets, indicated that personal intake of green vegetables was *overestimated by a factor of three!*

So even if you *think* you eat a reasonable quantity of green vegetables, you're probably wrong! Want proof? Write down everything you eat for two weeks.

No cheating! Include the meals you had in restaurants and at parties. They count, too.

When you add the items in the fresh green veggie column, you may be disappointed.

Of course, scientists generally concede that it's impossible to measure anything without creating change in some way, and this is a prime example of that phenomenon. If your two-week diary indicates substantial servings of green vegetables several times daily, I'll still take credit for inspiring your good dietary habits, thank you.

You don't need me to tell you to eat your greens. You can probably still hear your mother's voice scolding you with exactly that message. But why? What's in them that's so important? And why choose grasses like kamut instead of the more familiar table vegetables?

We cannot live by bread alone, and particularly not by sawdust bread.

4

NUTRIENTS AND DISEASE

The Medical Community Slowly Wakes Up

Cited in the Literature

Those of us who have followed the so-called natural health food movement over the years have been aware that there is a strong link between certain food, nutrients and immunity, and between immunity and degenerative disease. Occasionally, even the mainstream press offered tidbits about these connections. But in the past few years, conservative mainstream medicine has begun to report the irrefutable findings of these associations by researchers all over the world. I'll quote from the abstracts of a few contemporary examples, the result of a quick search through a computer database that you can access yourself through the library of almost any major university.

EXTRA!!! The Times EXTRA!!!

Vegetables Contain Nutrients
Protective Against Cancer

~ *Science*, April 1994.[7]

Many individuals in the United States have suboptimal diets; the potential for disease prevention can be improved by nutrition.

A general statement, but one that was considered controversial only a decade ago.

~ *Cancer*, August 1993.[8]

Diet may be an important factor in the cause and prevention of cancer.

The scientific community is catching on to the specifics!

~ *Journal of the American College of Nutrition*, April 1993.[9]

Consuming elevated levels of antioxidants such as ascorbate, carotenoids, and tocopherol is associated with delayed development of various forms of cataract.

Note that vitamins are referred to as antioxidants. Is this mainstream medicine's way of saving face — as though it can't admit that the health industry has been right about vitamins all along?

~ *European Journal of Cancer Prevention*, July 1993.[10]

The processed juice made from a number of vegetables significantly inhibits cancer-causing nitrosamines from forming.

The doctors learn about vegetable juicing. What took them so long? But note that this report appeared in a European journal. Is this too "far out" for our own New England Journal of Medicine?

~ *Institut scientifique et technique de la nutrition et de l'alimentation*, Paris, January 1993.[11]

The consumption of foodstuffs with a high vitamin C content should be recommended as it increases the bioavailability of nutritional iron.

So what else is new?

~ *Journal of Nutrition*, June 1993.[12]

Ascorbic acid ingested as cooked broccoli or as synthetic ascorbic acid seems to be equally bioavailable.

An admission that a synthetic vitamin can equal a food, in at least one respect! Yes, they're catching on.

~ *American Journal of Clinical Nutrition*, June 1994.[13]

Plasma concentrations of beta-carotene were increased after supplementation with beta-carotene capsules for ten to twenty days. Addition of cooked carrots to the diet resulted in no significant change in plasma beta-carotene.

Again, this demonstrates the superiority of a supplement over a cooked food. They're not fully informed yet, though. Young green grass shoots are replete with bioavailable carotenes that are easily digested, absorbed, transported, and utilized.

~ Department of Medicine/Nutrition, Medical College of Georgia, 1992.[14]

Micronutrients, including vitamins C, E, A, beta-carotene and other carotenoids should be the cornerstone of a cancer-prevention strategy starting now.

So it appears that young grass shoots are just what the doctor is ordering!

~ *Endocrinology*, May 1993.[15]

Vitamin D might exert beneficial actions on prostate cancer risk.
Is prostate cancer increasing because we are told to stay out of the sun, leaving us without a valuable source of vitamin D? Why not offer advice on getting antioxidants to help deal with the sun appropriately? See pages 50-54.

~ *Orvosi Hetilap* [Hungarian Medical Journal], March 1993.[16]

A protective effect of vitamins A, C, and E are shown for gastroduodenal ulcer and gastric cancer.
We know where to find a good free-radical demolition team, don't we? Young kamut grass, of course!

~ *Clinical Investigator*, January 1993.[17]

Cardiovascular mortality reveals an increased risk of ischemic heart disease and stroke at initially low levels of carotene and/or vitamin C. Low levels of both carotene and vitamin C increase the risk further in the case of stroke.
Again: young green shoots contain these very nutrients.

~ *Diseases of the Colon and Rectum*, March 1993.[18]

Antioxidant vitamins (A, C, and E) lower the recurrence rate of adenomas (benign tumors) of the large bowel leading to colorectal cancer and can be proposed as preventive agents.
Colorectal cancer is a totally preventable disease. Audrey Hepburn did not have to die in her sixties!

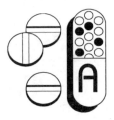  

~ *Cancer Causes and Control*, January 1993.[19]

A Canadian national breast screening study of 56,837 showed that those at the uppermost level of dietary fiber intake had a 30 percent reduction in risk of breast cancer relative to that for women at the lowest level. Inverse associations were observed in association with consumption of vegetables rich in vitamins A and C.

Breast cancer is another totally preventable disease, now affecting one in eight women. Lignans in green plant foods are particularly protective. Another plus for kamut concentrate as a supplement.

~ *Cancer Research*, February 1993.[20]

In a study of 25,802 adults, levels of all individual carotenoids, particularly beta-carotene, were lower among those who developed oral and pharyngeal cancer.

There it is again — carotenoids!

~ *Cancer Research*, February 1993.[21]

Food intake was assessed in 41,837 Iowa women, aged 55 to 69. High intakes of green leafy vegetables were associated with an approximate *halving* of risk. A lower lung cancer risk was also seen for high vitamin C vegetables, carrots, and broccoli, and for the nutrients beta-carotene and vitamin C.

So why didn't we listen to Mom when she said, "Eat your vegetables!"

~ *Chest*, January 1993.[22]

Nutrient treatment of smokers, ex-smokers, and others at risk represents a viable option. Agents proved effective in both laboratory models and humans include vitamin A and the carotenoids.

How much suffering could have been spared if this information had been more widely dispensed over the years!

~ *Chest*, January 1993.[23]

Prevention entails using specific agents to suppress carcinogenesis and thereby prevent the development of cancer. Beta-carotene, natural vitamin A, and the retinoids may be effective chemopreventive agents. However, ongoing administration of such agents may be required to prevent the development of cancer.
That's why I take my kamut "green power" every day.

~ *International Journal of Vitamin Nutritional Research,"* 1993.[24]

The overall composition, vitamin, carotene and mineral contents of the leaves of four wild plants used as vegetables were determined. The results are within the range of data published for domesticated leafy vegetables. The leaves revealed to be good sources of iron and calcium. Special attention is drawn to their contents of mucilages.
Yet another star for leafy greens.

~ *European Journal of Clinical Nutrition*, January 1993.[25]

Children can eat enough leafy vegetables to meet a day's need of vitamin A precursors. But feasibility of feeding children enough green leafy vegetables at home on a regular basis needs further study.
Yes, it surely does need further study, because I don't know how to get children to eat enough leafy greens, do you? That's why we add kamut's green power to my grandchildren's diet — dissolved in watered-down juice.

~ Martin Luther Universitat, Bundesrepublik Deutschland, *Nahrung*, 1993.[26]

The criterion for the nutrient value of vegetables is their specific contribution to nutrition, whereby particular emphasis must be placed on vitamin C, carotene, raw fiber, calcium and iron. But quality changes take place as a result of storage, temperature, and time.

The very reason I would never be without my concentrated green juice, dependable for standardized quality!

~ *European Journal of Clinical Nutrition,* 1993.[27]

A group of women, placed on an uncooked vegan diet including wheat grass juice, increased their energy and nutrient intake. In spite of the increased caloric intake, they lost 9 percent of their body weight.

I propose the next diet fad: green kamut concentrate! Wouldn't you like to see that sweeping the country?

~ Department of Chemistry, Helsinki, Finland, 1994.[28]

Dietary use of antioxidants provides protection against acute toxicity.

In addition to antioxidants, chlorophyll is a great detoxifier, as explained later.

~ *New England Journal of Medicine,* 1995.[29]

A study by several Boston area research centers claims that optimal intakes of folic acid, vitamins B12 and B6, may be essential for the prevention of heart disease. The study involved 1,041 patients.

Yes, they are waking up. Nutrition-aware researchers (including me!) have been writing articles about this very benefit for more than a decade.

You can't help but notice the prominence of vitamins A and C and the carotenoids in the list of nutrients reported in these prestigious journals. Although Grandma didn't identify the specific nutrients, she knew the foods that contained them had a protective effect against disease. That's why she told us to eat our vegetables. Do you think we would be more compliant if our *doctors* told us what to eat?

Although I've quoted studies published recently, the idea that diet and nutrition have an important influence on health is age-old.

> The link between diet and disease was mentioned in Chinese medical writings in the twelfth century.[30]

America's Eating Habits

People eat foods, not single nutrients. Confounding effects are associated with eating habits — age, body fat distribution, physical activity, alcohol consumption, tobacco use, psychosocial stress, and foods eaten together. The last is a highly significant factor too often overlooked.

For example: Among the dynamics influencing calcium absorption are foods ingested in the same meal. Foods that *enhance* the absorption of calcium are fatty fish, eggs, butter, and liver. Foods that *diminish* the absorption of calcium are sodas, unleavened bread, and milk. So two people with the same calcium intake will absorb different amounts of calcium if one is enjoying scrambled eggs for breakfast, salmon salad for lunch, and liver and onions for dinner, while the other has cornflakes and milk for breakfast, drinks a diet Coke midmorning, has a hamburger and another Coke for lunch, yet another soft drink in the afternoon and/or with a dinner of beef stew, and consumes pita pouches or unleavened bread on a daily basis. (The Coke and pita bread contain phosphorus, a calcium antagonist.) Both diets offer calcium of equal quantity, but the amount of calcium utilized can be very different. Nonetheless, we can get an overview from looking at the statistics. Two studies attempt to describe America's eating habits.

~ *Missouri Medical Journal*, 1993
Only 28 percent of Americans reported consuming five servings of fruits and vegetables per day, which is the World Health Organization goal for *HEALTH FOR ALL by the Year 2000*. Older females had the highest rate of consuming five servings per day, while young males had the lowest rate.[31]

~ *Journal of the American College of Nutrition,* June 1994
Nearly one-third of the food consumed by Americans is energy-dense and nutrient-poor, and comes from the food category called "other," which includes fats, sweets, and alcohol.[32]

They're calling in my loan. They said I'm depreciating faster than my car.

What About the Children?

In New York State, 1,797 second and fifth grade elementary children were studied. Here are some of the results:

> ~ 40 percent do not eat vegetables, except for pota-toes or tomato sauce.
> ~ 20 percent do not eat fruit.
> ~ Fifth graders eat significantly more snack foods and are more likely to skip breakfast than second graders.
> ~ Boys have lower food-group pattern scores than girls.
> ~ Children of lower socioeconomic status have less diverse diets but eat fewer snack foods than children of higher socioeconomic status.
> ~ Children with single parents are more likely to skip breakfast and eat fewer vegetables than those with two parents.
> ~ Children with mothers employed outside the home have less diverse diets than those with moth ers at home.[33]

An increasing social advantage is associated with decreasing dental health problems. Syrup medicines and, in particular, sweetened antibiotic liquids independently increase the risk that a child will have a number of carious lesions, especially if taken frequently.[34]

Over 80 percent of children aged two to five years consume more total fat, saturated fat, and cholesterol than is recommended. The major source of total fat and saturated fats is milk.[36]

Children's medicines —
Causing or curing?

IMPORTANT!

Our Children and Their Food

~ 36 percent of our children eat at least four different types of junk snack foods.

~ Skipping breakfast results in a very specific cognitive deficit.[35]

~ 16 percent of fifth graders do not eat breakfast.

~ Milk products, the most allergenic food we consume, are part of the diet of 82.7 percent of teenagers one or more times a day.

~ Inflammatory bowel disease is one of the most significant chronic diseases afflicting children and adolescents.

What About Our Teenagers?

Do adolescents fare any better? As we know, adolescence is an intense anabolic period when the requirement for *all* nutrients is increased. Note the following:

> ~ Meat, which contributes to calcium deficiencies and carcinogenic environmental hormone intake, is consumed by 66.3 percent of teens.
> ~ 77.7 percent eat biscuits, sweets, and chocolates.
> ~ Only 44.2 percent eat green and root vegetables between 4 and 6 times a week.
> ~ 73.4 percent eat out of the house between four and six times a week.[37]
> ~ The dietary practices of adolescent athletes fail to meet the energy requirements for high performance and may also threaten their well-being. Students fail to recognize nutritional practices critical to the demands of athletics.
> ~ Physically active adolescents have energy requirements far beyond those who are sedentary. Unfortunately, interest in nutrition is rare unless connected to improved performance.

Young people are increasingly taking a "grazing" approach to eating, rather than eating proper meals — with undesirable nutritional and social consequences. Television watching has been reported by the Division of Adolescent/Young Adult Medicine, Children's Hospital, Harvard Medical School to be associated with obesity, resting energy expenditure, and lower physical activity among both children and adolescents. (We didn't need a study to tell us that!)

The National Institutes of Health, Bethesda, Maryland confirms that both obesity and high blood cholesterol levels in our children are higher than optimal and that the benefit of reducing the prevalence of these conditions in childhood will be realized in adulthood.

Imagine the effect of making kamut juice concentrate mandatory for all children as a school snack, just as our cities offered *milk* to every child when I was in grade school!

Someone was on the right track with the introduction of the United States Department of Agriculture's 1991 *Eating Right Pyramid* food guide. At long last, the worthless Four Foods Group concept would be replaced with a program that would help to educate people about which foods were health promoting and which weren't.

But guess what? The guide was withdrawn in response to pressure from meat and dairy producers — just the latest in a long series of industry attempts to influence federal dietary recommendations. Such influence began when diet-related health problems in the United States shifted in prevalence from nutrient deficiencies to chronic diseases, and dietary advice shifted from "eat more" to "eat less."

The pyramid controversy focuses attention on the conflict between federal protection of the rights of food lobbyists to act in their own self-interest and federal responsibility to promote the nutritional health of the public.

Since 1977, for example, under pressure from meat producers, federal dietary advice has evolved from "decrease consumption of meat" to "have two or three daily servings." Thus, this recent incident also highlights the inherent conflict of interest in the Department of Agriculture's dual mandates to promote U.S. agricultural products and to advise the public about healthy food choices.[38]

Historical Development of Vegetarianism
In 1838, the American Health Convention passed resolutions which endorsed a vegetable diet as safe for all and also condemned the use of medicines. But such ideas were "too advanced for the time" and were described as nonsense. While many middle-class intellectuals eagerly embraced diet reform, the mass of Americans were not disturbed, and considered the reformers to be cranks.

A report cited in the *American Journal of Clinical Nutrition* in 1994 describes the history of vegetarianism as follows:

> Vegetarianism pursued for reasons of physical health is a recent practice historically. Before the nineteenth century, avoidance of animal food was justified with moral and metaphysical arguments. During the early 1800s, however, an intensified desire for improved health combined with the ascendancy of science to a position of cultural authority helped to promote the formulation of physiological arguments for vegetarianism. Theories of the nutritional superiority of a vegetable diet were nevertheless shaped by moral convictions, giving vegetarian spokesmen such as Sylvester Graham and John Harvey Kellogg the appearance of being dietary fanatics. Only as nutritional science expanded from the mid-twentieth century onward did vegetarianism acquire general recognition as a healthful dietary alternative. But because that alternative is still often selected for moral or other nonscientific reasons, nutritional education of vegetarians remains an essential activity.[39]

5

VITAMINS

Why Greens Are Your Best Source

Vitamin C
If I had named the vitamins, I would have named vitamin C
the *first* vitamin and called it *Vitamin A*. Vitamin C is
necessary for *any* biological process requiring nutrients to
pass through cell membranes — and that includes just about
all metabolic actions we know about! Perhaps the most
important role of vitamin C is to help prevent the proteins
in the cell membranes from being damaged by oxidation.

Proteins are large, complex, and specialized molecules that
control the functioning of cell membranes. They let nutrients in, send waste products out, and block entry to toxic or
viral invaders. Cell membrane proteins are responsible for
immune responses on the cellular level, *where it counts*.
When a protein is oxidized, it's split apart and destroyed.

Have you noticed what happens to white
cotton socks when you wash them with
too much bleach? Bleach is a powerful
oxidizing agent.

The proteins that hold the cotton together are damaged (along with the stains), and the socks are likely to fall apart after a few washings. You don't want that to happen to your cell membranes!

"Free radical" is a term you hear a lot when oxidation and vitamins are discussed. This is a bit of chemistry jargon that simply means a charged molecule or molecular fragment (a group of atoms) having an electrical charge. The charge is usually the result of one or more oxygen atoms not fully bonded to the rest of the molecule. It's available to split off and oxidize something else, like a cell membrane protein.

Elson Haas, M.D., in his excellent book, *Staying Healthy with Nutrition,* says:

> The free-radical theory, currently the most accepted aging hypothesis, offers an explanation of the basis of degenerative disease.[40]

I have more and more trouble remembering people's names. I sure hope I'm not getting what's-his-name's disease!

Vitamin C works by rushing in first and neutralizing the free radicals before they can do their damage to cell membranes and other tissues. Interestingly, humans and a few other primates (apes, chimpanzees) are the only mammals that don't synthesize their own Vitamin C.[41] We have three of the four necessary enzymes, and it's speculated that we lost the fourth (L-gulonolactone oxidase) through evolutionary misfortune. No problem, if you're living in the woods and eating everything fresh off the vine. In fact, that's probably why we have a sweet tooth — to seek out the freshest, most vitamin-rich greens and fruits.

Put that same primate in the convenience store instead of the jungle, and you get a different result. Lots of sugar, no vitamins and degenerating health.

The missing enzyme is necessary for the last step in the transfer of sugar to ascorbic acid. Again, no problem if you eat fruits and vegetables with high ascorbate content. Too bad we can't make vitamin C from sugar, as other animals do. Actually, the commercially manufactured ascorbates are made from glucose!

How much vitamin C do you need? Goats, who weigh about the same as humans, synthesize around 50 grams of vitamin C per day. That's *50,000 milligrams!* (The RDA, the recommended dietary allowance prepared by the Food and Nutrition Board of the National Academy of Sciences, says we need only 60 milligrams.) I take a few thousand, more when I'm forced to go for a day without greens.

The best way to get vitamin C is from food. Bioflavonoids, associated with vitamin C in nature, are not fully understood. Synthetic ascorbic acid, chemically similar to vitamin

C (or almost similar — sometimes it's a molecular "mirror image") lacks cofactors like bioflavonoids. The better choice would be supplements that include bioflavonoids.

A complete plant leaf is guaranteed to have all the cofactors.

See what I'm getting at? Because there's so much we *don't know* about how vitamins work, the best source of vitamins will always be the whole foods that contain them. Not necessarily because the vitamins themselves are any different, but because other substances, commonly found with the vitamins, probably have important but as yet unidentified roles to play.

Vitamin C is found in citrus, berries, cantaloupes, tomatoes, potatoes, papaya, parsley, peppers, cabbage, and *green vegetables*. But it is an extremely fragile nutrient. After cutting an orange in half, 7 percent of the vitamin C dissipates in half an hour.

Vitamin C antagonists are air pollution, industrial toxins, smoking, alcohol, aspirin, antidepressants, diuretics, improper food preparation, antibiotics, and high fever.

Vitamin A and Beta-carotene

About ten million children the world over are deficient in vitamin A.

> Over one million suffer from varying degrees of visual impairment every year.

"This problem is likely to be seriously magnified by the year 2000," reads a report in the *Journal of Ophthalmic Nursing Technology*. "The scope of the problem is immense, and the need to address it is urgent, representing one of the greatest failures in global public health planning," conclude the researchers, who suggest that vitamin A deficiency can be eradicated with *five cents and a vegetable garden!*[42] While most of the vitamin A deficiency problems have been done in third world countries, the effects are seen here, too.

Much of the research cited earlier links vitamin A intake to cancer resistance. Patrick Quillin sums it up well in his book *Healing Nutrients*:

> So important is vitamin A in preventing cancer that serum levels of vitamin A are like a crystal ball in predicting who will get cancer: the lower the level of serum vitamin A and/or beta-carotene the greater the risk for cancer.[43]

Like vitamin C, vitamin A is a very powerful antioxidant but seems to be required in much smaller doses. Because it's fat-soluble rather than water-soluble, it's easily stored in body tissue. In extremely large doses vitamin A can be toxic. Beta-carotene, however, is one of the water-soluble precursors of vitamin A — a nontoxic, powerful antioxidant in its own right.

Carotenes are the forms found in plants. Most body cells can convert carotenes to vitamin A as required. The carotene in leafy greens is converted to vitamin A about twice as efficiently as the carotene in carrots and other root vegetables.[44]

Beta-carotene is just one part of the carotene complex. It happens to be the easiest to synthesize and the one that has been used extensively in research. But it is not necessarily the most important. Alpha-carotene, for example, is far more effective than beta-carotene in combating some types of human cancer cells. A report in the *Journal of the National Cancer Institute* states:

We found that natural carotene extracted from palm oil suppresses the proliferation of various human malignant tumor cells....It was about ten times more inhibitory than beta-carotene.[45]

This is yet another good argument for obtaining most of your vitamins from food or concentrated food supplements, sources that are far more likely to contain the full range of carotenes, as opposed to synthetic extractions.

The reason carotenes are such good antioxidants is related to the process of photosynthesis. Plants that convert solar energy to chemical energy must deal with charged particles everywhere. The sun's rays, in the form of photons, send electrons bouncing in every direction. With the aid of chlorophyll (and hundreds of other chemicals) the result is the manufacture of carbohydrates and the release of oxygen. One would think that this reaction, going on right there inside the plant leaf cells, would be extremely damaging to the surrounding cells. These cells do require powerful protection from the onslaught of stray charges, and they get it from *carotenes*. That explains why the darkest, greenest plants — the ones with the most chlorophyll — tend to have such high concentrations of carotenes.

We also know that the *fastest-growing* leaves have the most chlorophyll. Guess which leaves grow the fastest? Cereal grasses, like the young kamut plant! In general, the darker the green vegetable, the higher the carotene concentration. The yellow-orange color is masked by the green pigment of the chlorophyll, but where there's lots of chlorophyll there's plenty of carotene. Of course, carrots are also full of carotene. A large raw carrot has about 11,000 IU of carotene. Many health practitioners are recommending 25,000 IU as part of a general-purpose optimum-nutrition regimen.

Beta-carotene vegetable foods include spinach, sweet potatoes, pumpkin, carrots, butternut and Hubbard squash, collard greens, dandelion greens, kale, turnip greens, beet greens, red peppers, Swiss chard, bok choy, mustard greens, tomatoes, and broccoli.

Vitamin A antagonists are air pollutants, exposure to glare or strong light, nitrate fertilizers, vitamin D deficiency, alcohol, coffee, cortisone, and mineral oil.

The B Vitamins

If you're on an *antibiotic* or you take aspirin or other drugs; if you happen to be *fasting* because your friend said it would make you feel so good; if, like most of the world, you're *dieting*; if you're under *stress* because of a traffic jam, deadline, or mother-in-law problems; or if, like just about *all* of us, a major portion of your food is cooked (have I missed anyone?), more than likely you are vitamin-B deficient!

Because they are water-soluble, B vitamins are not usually stored in your body — it isn't difficult to subject yourself to short-term deficiencies, especially since most B vitamins are easily damaged or destroyed by cooking. Some are produced by intestinal bacteria, so if you're taking antibiotics, your normal supply may be cut off. (Antibiotics are not selective — they annihilate all bacteria, including the good-guy flora.) Nutrient depletion for any reason (dieting, fasting) curtails the supply, and stress is a major B vitamin destroyer.

> Although B vitamins are usually bundled together in foods, the separate B vitamins can have very distinct functions.

Some B vitamins play a major role in amino acid and enzyme production. Others participate in immune function and the regulation of certain toxins. B6, for example, is required to metabolize an amino acid that is responsible for initiating damage to arterial walls, causing atherosclerosis and coronary disease later on. For this reason, B6 deficiency may prove to be a major factor in heart disease, our number one killer. Folic acid (another B vitamin, named for the foliage in which it is found) helps to regulate new cell growth and is related to certain immune responses. B vitamins are also linked to proper functioning and health of the nervous system, skin, hair, and eyes.

Vitamins B3 and B6 deficiencies are linked to arthritis; folic acid to anemia; B6 and B12 to carpal tunnel syndrome; B6 with PMS (because it's a cofactor in the synthesis of serotonin); and niacin (B3) with high cholesterol and insulin intolerance. Unfortunately, signs of marginal B vitamin deficiencies are difficult to detect.

Isadore Rosenfeld, M.D., although he is on the conservative side when it comes to getting nutrients from tablets, capsules, and powders, suggests that the quickest and best way to correct a vitamin B deficiency is with supplements![46] Whole grains, rich in B vitamins, are consumed by fewer than 20 percent of Americans. Over the years, I have found that B-complex needs could be met with concentrated food-type supplements. A half century ago, brewer's yeast was the only game in town. Today, I have choices. To paraphrase Cher, *I choose kamut!*

Vitamin B foods include whole grains, fish, leafy greens, yogurt, seeds, *green vegetables*, yeast, nuts, and soybeans.

Vitamin B antagonists are alcohol, cooking, storage, refined flour, high temperatures, stress, caffeine, aspirin, environmental pollution, and the Pill.

Vitamin D

Vitamin D is a fat-soluble vitamin that closely resembles cholesterol. This is one vitamin that you don't necessarily need to ingest — you can synthesize it from sunlight! When skin is exposed to the ultraviolet light that travels from the sun, a precursor of vitamin D is manufactured from a form of cholesterol — one of many important functions of cholesterol. This substance is then converted into the active form of vitamin D by your liver and kidneys. It's a fortunate person who can spend enough time outside to take advantage of the body's capability to do this.

Vitamin D is important in regulating both the calcium and phosphorous flow into and out of bone structure, as well as for numerous other functions related to your heart, nervous system, and blood. Because it's produced in one part of your body (skin) and has a controlling effect on other tissues (bone), vitamin D can be thought of as more hormone-like than vitamin-like. The converted form of vitamin D (1-25-dihydroxycholecalciferol or $1,25[OH]_2D$) is closely related chemically to estrogen and cortisone.

> Unfortunately, the practice of regular sunbathing has been drastically curtailed by the prevailing "wisdom" that sunlight is bad for you. Nothing could be further from the truth!

Skin cancer is more common than ovarian, cervical, central nervous system cancer, or leukemia. Melanoma (skin cancer) is increasing faster than any other cancer in the United States and all over the world. "And you're still going to say it's okay to sunbathe?" I hear you ask, incredulously. Well, think about this trend again. Certainly Americans are not spending any more time in the sun than they did a generation ago. Just the opposite is true. We are likely to spend more time in the glow of the TV set! As a nation, our lifestyle has become ever more sedentary — enclosed by buildings and protected from the sun.

If the correlation between sun exposure and malignant skin cancer were that simple, surely skin cancer rates would be dropping!

Note the following research:

~ *International Journal of Epidemiology.*
Lack of exposure to ultraviolet sunlight can increase the prevalence of vitamin D deficiency and may place some women at higher risk of breast cancer. A statistically significant negative association was found between breast cancer incidence rates and total sunlight levels.[47]

~ *Preventive Medicine.*
Regular sunning could prevent far more breast cancer fatalities than the skin cancer fatalities it would cause.[48]

~ *Cancer.*
Ultraviolet radiation may protect against clinical prostate cancer.[49]

~ *Anticancer Research.*
Mortality rates from prostate cancer in the United States are inversely correlated with ultraviolet radiation, the principal source of Vitamin D.[50]

Still want to hide from the sun? Not me! On the other hand, the risk of damaging your skin is very real, and the increase of serious skin cancer should not be ignored just because other life-threatening risks may be reduced. While vitamin D may explain part of the beneficial effect of sunlight, we still need to understand what is failing in our skin that makes us so much more susceptible to skin cancer than our ancestors were a few generations ago.

The answer? Antioxidants, which are themselves easily oxidized because they can have a sacrificial role, intercepting reactive particles (the free radicals, explained in more detail later) before they damage vital lipids and proteins. Not only is skin exposed directly to the mutagenic solar radiation (that feels so good), but skin also has a very high rate of regeneration for healing wounds and for replacing worn-out tissue.

Viewed on a small enough scale, it's correct to say that everyone always has skin cancer. The DNA in skin cells is constantly subject to damage from high-energy sunlight. When this happens, the cell's "programming" to regenerate sometimes takes over. Uncontrolled cell proliferation, fortunately, is the exception rather than the rule. Cells can communicate with each other via an astoundingly complex chemical language, and the programs in the undamaged DNA can identify abnormal cell propagation and shut down the new cell line. When your cells are bathed in antioxidants, this is happening during every minute of sun exposure.

One interesting study of human melanoma in Moscow concluded that consumption of greens significantly decreased the risk of melanoma.[51]

This is certainly not a surprising result, and it suggests that the real problem with sunlight and skin cancer is nutritional. Keep your skin well supplied with antioxidants and the risk is minimized.

It cannot be overemphasized that the best natural sources of antioxidants are the greenest and fastest-growing plants. This makes sense in light of what we learned earlier about what goes on inside the leaf of a plant. Extremely high-energy solar radiation is being used to drive the photosynthesis reaction, which in some ways is the reverse of oxidation. It's critical to keep the newly formed products of the plant — the carbohydrates and free oxygen — from bouncing right back in the other direction. It's also important to control the unwanted effects of all that high-energy radiation and those highly charged particles associated with the photosynthesis process.

New material is being synthesized at a phenomenal rate as the plant grows. Again, with such powerful electrical and chemical forces at work, a large supply of antioxidants is necessary to keep these processes from causing major damage. Free radicals can be likened to a single checker jumping over a row of other checker pieces, wiping them out as it crosses the entire board.

Is it any wonder that the immature cereal grasses, among the fastest-growing of edible plants, are also a superb source of antioxidants?

Of course you're going to be careful not to overdo sun bathing — burning can't be good for you. But a healthy tan is exactly that— *healthy!* Someday soon, the high priests of medical wisdom will understand how important sunshine is for health.

The implication for sun worshipers is clear: Eat lots of fresh green plants, with emphasis on the young sprouted shoots.

Glass, clothing, and smog block the ultraviolet rays that create vitamin D, but clouds do not. Blood values of vitamin D are determined more by previous exposure to summer sunlight than by dietary intake of vitamin D! A test of hospital inpatients revealed that their levels of vitamin D increased during the summer, even though they never ventured outdoors. It was assumed that this was caused by a natural increase in the vitamin D content of some foods during this time. Vegetables grown between May and July offer bone-health properties. Another plus for kamut juice concentrate!

Vitamin D deficiency is the most common nutritional deficiency in Crohn's disease, which is fast replacing ulcers as a common digestive disorder among Americans.

Vitamin D is found in sunlight, fish liver oils, eggs, brewer's yeast, and shrimp. Mushrooms, avocados, and dark green leafy vegetables are good plant sources.

Vitamin D antagonists are cortisone, inadequate exposure to sunlight, mineral oil, laxatives, antacids, and aflatoxins.

Vitamin E

Vitamin E has been credited with everything from improved sex hormone production to prevention of hair loss. We do know that vitamin E, another fat-soluble antioxidant, offers protection against heart and vascular disease and against a wide range of toxic substances.

One problem with vitamin E supplements, however, is that the oil-based products from which it is extracted are usually very unstable. Light, heat, air, and age render the oil rancid and the value of the supplement questionable. This seems like a paradox at first: Why should a preparation so rich in vitamin E, a powerful antioxidant, be so easily oxidized itself? The answer is that vitamin E works *because* it's so easily oxidized. It presents itself as an easier target for the free radicals, thereby protecting vital proteins and lipids from damage.

What this means to you and me is that we should not rely completely on supplemental forms of vitamin E. As usual, it's much better to get it from food.

For reasons we do not yet understand, certain nutrients are absorbed better (and therefore less is required) when we consume them in a whole, natural context. Vitamin E may be one of these nutrients, and the very reason that the germ of the wheat has been touted as a good source for vitamin E. According to Dr. Bernard Jensen, we must eat vitamin E in its raw state to get its full benefit. He recommends chewing on raw wheat kernels.[52] I suggest that the kernels be sprouted first.

If you care to bother, you can do as I do. I soak a handful of kamut grains every day in a glass jar, rinse them twice a day, harvest them on day three when they have sprouted just enough to be sweet (but not too sweet), and toss them into that day's salad. I know that the vitamin E content in the germ, protected by the tough-coated seed, will not be rancid.

This cornucopia not only supplies me with vitamin E but also with a small amount of good protein from the endosperm of the grain, and even fiber from its outer covering. But I caution you — it is "another mouth to feed": The seeds in the jars have to be rinsed twice a day. I think it's worth the effort. Did I mention the nutty, chewy contribution of the sprouted kamut grains to your salad?

Exercisers, take note: Vitamin E decreases damage occurring from aerobic activity!

Vitamin E is found in whole grains, eggs, avocados, sweet potatoes, *green leafy vegetables*, asparagus, and broccoli.

Vitamin E antagonists are oxidizing agents, food processing, rancid fats and oils, inorganic iron, the Pill, mineral oil, and chlorine (as in tap water).

6

MINERALS

How They Relate to Kamut Grass

In the interest of brevity (and not to bore you), I have selected only a few minerals for discussion. But it should be enough to give you insight into the importance of consuming green leafy vegetables and young grass plants.

Potassium and Sodium
Nearly everyone who consumes the typical American diet ingests too much sodium and is deficient in potassium. The trouble is, our bodies are designed to thrive in a world where sodium is scarce and potassium abundant. And for good reason! If you look at the potassium and sodium content of most *natural* foods, you'll find they have plenty of potassium but little sodium. Consequently, a complex control system, good at hoarding sodium and disposing of surplus potassium, is part of our biological makeup. Now study the potassium and sodium content of *processed* foods, and you will find the relationship between these two minerals reversed!

POTASSIUM AND SODIUM CONTENT
OF SOME COMMON FOODS
BEFORE AND AFTER PROCESSING
(milligrams/100 grams food)

	POTASSIUM	SODIUM
Flour, whole	360	3
White bread	100	540
Pork, uncooked	270	65
Bacon, uncooked	250	1400
Beef, uncooked	280	55
Corned beef	140	950
Haddock, uncooked	300	120
Haddock, smoked	190	790
Cabbage, uncooked	390	7
Cabbage, boiled	130	230
Horseradish, raw	564	8
Horseradish, prepared	290	200
Asparagus, raw	310	2
Asparagus, canned	250	200
Peas, uncooked	340	1
Peas, canned	130	230
(served with ½ oz salted butter)	99	374

Adapted from: Kamen, B. *Everything You Always Wanted to Know About Potassium But Were Too Tired to Ask* (Novato, CA: Nutrition Encounter, 1992), p 19.

Why is this ratio reversal important? Back to cell membranes! While complicated proteins *control* the membrane, sodium and potassium *power* them. Each cell membrane operates a kind of electrochemical machine called the *sodium-potassium pump*. No one knows *exactly* what gives the membrane the energy to pump all those  ions back and forth, but we do know that by using food energy to push charged particles of potassium into the cell and sodium out, an electrical charge is maintained. Building this electrical potential is like charging a battery. Your cell power can now be used by thousands of different cell membrane proteins to accomplish tasks that require energy.

As often as a thousand times a second, sodium and potassium ions exchange places to inform your brain about size, distance, patterns, color. If potassium is in short supply — or if the ratio of potassium to sodium is too low — your cell membranes become tired. Additional results of an unbalanced sodium/potassium ratio are:

~ compromised cell nourishment
~ reduced immune function
~ lowered resistance to toxins

Solution? Make sure your diet includes lots of potassium. Most authorities recommend at least three times as much potassium as sodium.[53] Some diets aimed at reducing high blood pressure suggest a ratio of 15 to 1.[54] But the typical American diet has about two to three times as much sodium as potassium — just the reverse of what you need!

Wheat grasses typically have 100 to 200 times as much potassium as sodium, a marvelous relationship! A typical wheat grass value is 400 to 450 milligrams of potassium per 100 grams — about half a percent of total weight. This is slightly more than you find in a banana, the traditional high-potassium food.

Organically grown foods are much higher in potassium than nonorganic foods — an obvious fact, but I repeat it only as affirmation and especially to remind *me* of how important food selection can be.

You can learn more about potassium and sodium from my book, *Everything You Always Wanted to Know About Potassium But Were Too Tired to Ask.* (Dr. Louis Pottkotter, a pediatrician in Texas, informed me that he gleaned information about nutrition from this potassium book that he was never taught in medical school!)

Good sources of potassium are fruit, especially bananas and oranges, fish, poultry, whole grains, vegetables, seaweed, nuts, seeds, legumes, yogurt, and blackstrap molasses.

Potassium antagonists are salt, cortisone, caffeine, stress, alcohol, laxatives, diuretics, sugar, high cholesterol levels, diarrhea, and excessive sweating.

IMPORTANT!

~ Even celery, a high sodium food, has more potassium than sodium.

~ The unnatural ratio of sodium to potassium is a major cause of fatigue.

~ Instead of salt, add basil, chives, dill, fennel, parsley, tarragon, or sesame seeds to fish.

~ To convert the weight of sodium to the weight of sodium chloride (salt), multiply by 2.54. If frank-furters contain 6 mg of sodium, they contain 15.24 mg of salt as sodium chloride.

~ Warning: Potassium supplementation should not be given to young children without consulting your physician.

HOW NOT TO DO IT

Food (Milligrams per 100 grams of food)	Sodium	Potassium
Apple pie	110.0	80.0
Bread, white	22.0	2.8
Cheese spreads	48.0	26.0
Cheese, American or cheddar type	30.0	2.1
Cold cuts	60.0	6.0
Corn bread	23.0	4.5
Cottage cheese	10.0	2.2
Dry cereal	4.3	3.0
Frankfurters	48.0	6.0
Green Olives	95.0	1.4
Pancakes	18.0	2.5
Peanut Butter	25.0	16.0
Peas, canned	10.0	1.2
Potato chips	42.0	11.0
Ritz Crackers	47.5	2.5
Salmon, canned	15.0	9.0
Saltines	48.0	3.0
Sauerkraut	32.0	3.5
Sponge Cake	9.0	3.0
Tuna, canned	42.0	14.0

Chart adapted from *Everything You Always Wanted to Know About Potassium But Were Too Tired to Ask*, B. Kamen (Novato, CA: Nutrition Encounter), p. 108.

HOW TO DO IT
Dark leafy greens!

Calcium

Bones are not what they appear to be. That is, they are not static structures, but living tissue.

A surprisingly large amount of calcium flows in and out of your bones every day!

If you didn't supply your body with a constant stream of calcium, along with magnesium, phosphorus, manganese, boron, and other minerals necessary to assimilate and construct bone material, your skeleton would crumble in just a few years. This is almost exactly what happens in osteoporosis, a disease afflicting an overwhelming percentage of women in North America. *In the time frame during which the number of American women doubled, the incidence of osteoporosis tripled!*

But why this risky exchange of calcium in and out of bones? The answer is that your body has a very low tolerance for variations in calcium concentrations in your blood. When more calcium is needed, it has to be available *immediately* because deficiencies lead to serious problems. Complex backup systems allow calcium to be grabbed from your bones to assure the maintenance of circulating calcium in your blood. Your bones are your calcium warehouse. It's vital to provide continuous replenishment.

Doesn't milk offer an adequate supply? No, it doesn't, despite what the medical journals would have you believe. The calcium in commercial cow's milk almost never has a positive effect on calcium assimilation for a number of reasons, among them:

> ~ an unbalanced phosphorus-to-calcium ratio
> ~ an altered fat profile
> ~ diminished enzymes (destroyed in the pasteurization process)

The fact is that cow's milk, although a superb food for calves, is a very poor food for humans. So here's yet another example of how food processing has a deleterious effect on our metabolism!

Where does all that calcium in milk come from, anyway? Do you see the farmer giving calcium pills to the cows? Mature cows never drink milk, and you won't find the farmer out in the field making them swallow calcium pills. Cows eat lots of *grass*. (So do I! In the form of young kamut juice concentrate!)

Good sources for calcium are *green vegetables*, broccoli, kale, bone marrow, blackstrap molasses, sardines (with bones), salmon, soybeans, tofu, and egg yolk.

Antagonists are high phosphorus, high protein diet (as in meat), stress, inactivity, deficiencies in magnesium, vitamin D and hydrochloric acid, drugs (including cortisone, antacids, laxatives, and diuretics), aluminum and tap water.

IMPORTANT!

Facts About Calcium

~ Sugar interferes with calcium absorption.

~ Salt encourages calcium excretion.

~ Tap water interferes with calcium absorption (caused by chlorine treatment).

~ Calcium may have an anticancer effect on your bowel by rendering the cells that line the colon less susceptible to malignant change.

~ When milk is pasteurized, the heat destroys certain enzymes, thereby causing the calcium to become insoluble.

~ Calcium must be consumed with synergistic nutrients to be effective.

When areas of the world with notably low dietary calcium (the South African Bantu region, Hong Kong, and Singapore) are compared with high calcium populations (Britain, Sweden, and the United Sates), it becomes obvious that high calcium intake alone is not associated with long-lasting healthy bones.

Figure 1

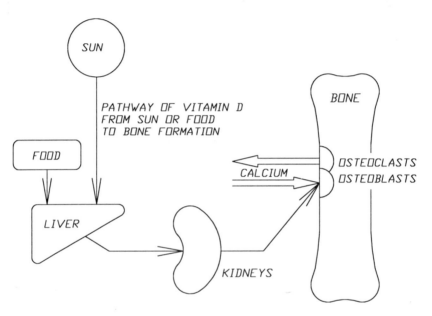

VITAMIN D FROM SOURCE TO CALCIUM

For calcium to be incorporated into your bone structure, vitamin D is a necessity. But this nutrient is provided by only a limited number of foods and by sunlight exposure.

Figure 2

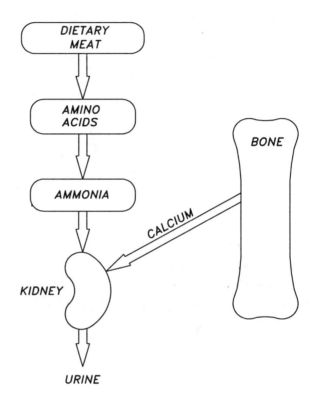

MEAT CONSUMPTION
AND CALCIUM EXCRETION

The metabolism of amino acid products in meat requires additional calcium to be pulled out of bone tissue.

Zinc and the Superoxide Dismutase Connection

Many steps are involved in the process of empowering you with energy — necessary dynamics to enable you to function — to blink, to wave your hand hello, to get up from a chair, to read this book, to show endearment to your significant other, or even to smile.

A small but significant amount of oxygen (only 1 to 2 percent) is used during the process of energy-infusing cell metabolism, and it is left as an *oxygen intermediate*. The problem is that an oxygen intermediate can do a great deal of damage to surrounding tissue.[55] Oxygen intermediates can contribute to many illnesses, including inflammatory diseases, toxic reactions, cancer, and aging disorders.

For those who want the biochemistry, the result can be a *superoxide anion*, one of the most common oxygen intermediates. Written as O_2^-, denoting two oxygen atoms with one extra electron for a negative charge, superoxide is a kind of "natural" free radical. Superoxide and other oxygen intermediates can be used to good advantage. Phagocytic cells, for example, use them to surround and destroy invading cells of an infection or parasite. (Phagocytes are blood cells that ingest foreign particles like bacteria.)

But every other cell in your body needs protection against superoxide, and the enzyme *superoxide dismutase* (SOD) is the primary line of defense. *SOD is an oxygen-intermediate scavenger.*

Of the 100,000 enzymes in your body, SOD ranks number five in total concentration. More to the point, the theory has been advanced by gerontologist Dr. Richard Cutter that *lifespan is directly controlled by SOD.*[56]

Zinc plays a key role in SOD metabolism. Without zinc:
~ membrane lipids would be oxidized, altering the essential internal and external cell structure
~ cellular proteins and DNA would be oxidized and damaged
~ enzymes would neutralize, shutting down cell respiration, essentially the same as killing the cell

SOD has been shown to block the oxidation of LDL cholesterol (the "bad" cholesterol). While cholesterol itself may not be as damaging as popular conception would have it, oxidized LDL cholesterol is a serious hazard.

Zinc itself competes for binding sites with other elements such as iron or copper, and the effect is to reduce the oxidation of these other metals.

Antioxidants work by neutralizing free radicals before they can damage cell membranes and other tissues.

If unchecked, free radicals can trigger a chain reaction of sorts, breaking apart larger molecules and leaving more dangerous free radicals in their wake. A molecule of beta-carotene can deactivate one single oxygen radical, and it gets consumed in the process. A molecule of vitamin C can fight off two free radicals. Some enzyme systems can deal with thousands of free radicals. SOD is one such enzyme, and zinc appears to be a key cofactor. But it can't work without other antioxidant enzymes (superoxide catalase and glutathione) waiting in the wings to clean up the electrochemical mess made by the powerful (but sloppy) SOD. One antioxidant is seldom sufficient; antioxidants do best in teams. No matter how much zinc you consume, you still need an adequate supply of vitamin E.

The big lesson we've learned in nutrition in the last few years centers on the interrelationships of nutrients. The amount of zinc in an agricultural product is closely tied to the amount in the soil that supported its growth. Needless to say, most of the food in the supermarket vegetable section has far less of this vital mineral than the corresponding wild plant. Even though this is a trace mineral — a healthy human body has only about two grams — it's safe to say that zinc deficiency is common. Zinc is also water-soluble, so it's easily carried away with cooking water.

Zinc is needed for the synthesis of over a hundred different enzymes and has a role in male sexual functions (providing a possible explanation for the folklore associated with oysters, which are very high in zinc).

Whole grains are a good source of zinc. Note, however, that most of the zinc is in the germ and bran. Refining grain into white flour typically removes about 80 percent of this mineral.

Sources of zinc are oysters, herring, organ meats, eggs, bones, brewer's yeast, legumes, nuts, paprika, whole grains, mushrooms, *leafy greens*.

Zinc antagonists are alcohol, excess calcium, cadmium, unleavened bread, refined grains, phytates, and oral contraceptives.

7

OTHER NUTRIENT CONNECTIONS

Why You Should Know About Them

Fats and Oils

Remember when television commercials bragged about *polyunsaturate*d vegetable oil and how good it was for your heart?

> What you may not remember is that those ads were challenged in court and had to be retracted.

There was virtually no evidence of any health benefits from that particular product. A polyunsaturated oil extracted and refined at high temperature, packed in a clear glass bottle, and stored for weeks or months on the supermarket shelf is going to be useless to your heart. In fact, it could be deadly. This is because unsaturated oils, while beneficial in their fresh and unaltered state, are very unstable. Heat, light, air, and time transform the molecular structure of these oils, or, more specifically, alter the fatty acids into forms that are *almost*, but not quite, identical. When these fatty acids are used for hormone production, or when they are incorporated into cell membranes, the result is not what nature intended. Horace, two thousand years ago, said, "You may drive out Nature with a pitchfork, yet she still will hurry back."

The same damage occurs with saturated fats. (Saturated fatty acids are usually solid at room temperature and are commonly referred to as fats. Unsaturated fatty acids are more likely to be liquid and are designated as oils.) Even though a saturated fat may be more stable than an unsaturated oil, any significant amount of heat causes oxidation with its resultant harm.

> **The consequence of eating a char-grilled burger is a major dose of free radicals set loose to destroy the proteins and lipids in your cell membranes.**

The cell corruption is minimized if you are well supplied with antioxidants like vitamins C, A, beta-carotene, E, and the minerals zinc and selenium, but it's far better to avoid that burger in the first place. *Your antioxidants already have plenty to do.*

Fats and oils, however, are vital to good health. As with so many other nutrients, the best way to get them is to eat the foods that contain them naturally. It's especially important to have EFAs, the essential fatty acids. As defined earlier, these are the ones that can't be synthesized by your body, hence the name *essential*. In the fat category, they include linoleic, linolenic, and arachidonic acids.

> **Kamut grass is particularly rich in linoleic acid and also contains some linolenic acid.**

Figure 3   Various Forms of Linoleic Acid, an 18-carbon Fatty Acid

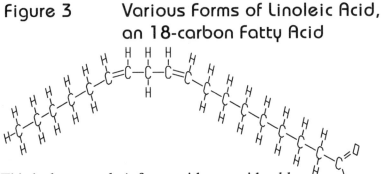

This is the natural *cis* form, with a considerable bend, keeping the oil in liquid form.

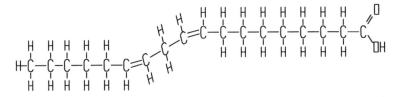

Exposure to heat, light, or time, straightens the carbon chain os the molecules can now stack up like cordwood, giving the oil a much higher melting point. A liquid *cis* fatty acid may become solid at the same temperature in the *trans* configuration.

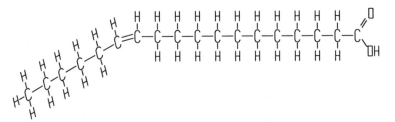

Partial hydrogenation can have somewhat the same effect. If one missing hydrogen atom is added back, the molecule straightens and the oil becomes more solid.

Enzymes

There's more about nutrition we *don't* know than we *do* know. One example is the "Grass Juice Factor," which has been designated as a property of grass juices, but eludes isolation and exact chemical classification. When laboratory animals are fed diets that include green grass juices, they are far healthier than those fed diets complete in all *known* nutrients. These experiments are described in more detail by Ronald L. Seibold in his book, *Cereal Grass, What's In It for You.*[57]

Nutritional science, however, isn't completely in the dark about food enzymes. An enzyme is a kind of catalyst — a substance that facilitates or promotes a chemical reaction but is not itself consumed by the reaction. Many chemicals perform as catalysts, but an enzyme is a *biological* catalyst — containing proteins of awesome molecular complexity constructed to help design DNA codes in a cell's nucleus.

Enzymes can be produced only by a living organism.

Virtually everything that goes on in our bodies is controlled by enzymes. Breathing, eating, sleeping, even thinking and emotions are based on enzyme reactions. Enzymes snap nutrients apart. Immune functions, hormone production, vitamin utilization, tissue replacement, and countermeasures against toxic substances rely on enzymes. Most of the codes in our DNA are dedicated to producing the right enzyme at the right time.

Enzymes can be classified into three functional groups:
- ~ metabolic enzymes, which control just about all metabolic reactions
- ~ digestive enzymes, which digest food
- ~ food enzymes, not manufactured in our bodies but contained in the foods we eat

Some of our glands and organs — our pancreas and liver, for example — could be called *enzyme factories*, producing a large variety of different enzymes for different purposes *on demand*. All this takes place well beyond our consciousness, which is probably a good thing. I know if I were to personally take manual control of my own liver I'd be in big trouble. I'd have a better chance of survival if I tried to land a 747! Why, then, are enzymes in foods important? Don't our bodies make all the digestive enzymes we need to deal with any natural food? Only recently have we begun to realize that the answer to the last question is *no*.

A food usually contains enzymes that are extremely useful for the digestion of that food. Enzymes found in most fruits convert starch into sugar. Seeds and nuts contain enzymes to digest fat. High-cellulose vegetables contain an enzyme to break down cellulose. And even lean meat contains an enzyme to digest protein.

Our stomach is also charged with completing these very tasks, but if the right enzyme is provided prepackaged with the food, then digestion is much easier and much more efficient.

Now the bad news. Temperatures that are higher than 118° Fahrenheit destroy most enzymes over time. Normal cooking temperatures kill them quickly. Hence the significant inherent superiority of raw food over cooked food. This advantage is dramatically demonstrated by comparing mother's milk to pasteurized milk. Mother's milk is enzyme-rich, serving as the only outside source of enzymes for the infant. Pasteurized cow's milk (aside from having the wrong mix of sugars and proteins for human infants) has had nearly all of its enzymes destroyed or distorted during pasteurization, homogenization, and storage — not to mention the enzyme-poor diet of the milk-factory cow. That's why breast-fed babies are healthier throughout life, and even smarter — they actually have more brain cells.

Honey is another example, except that in this case the enzymes originate in the flowers from which the bees draw nectar to make the honey. These enzymes are very useful to humans in the digestion of starches (e.g., the bread on which the honey might be spread).

Commercial honey, even if it's not pasteurized, may be made by bees that are fed sugar for at least part of the season, cutting off the enzyme supply.

The lesson to be learned from our quick study of enzymes in food is that it's vital to eat a large fraction of our food raw and fresh. This is easy to say, difficult to do. Unless we are willing to throw away our modern culture and walk back into the woods where everything is fresh and raw, we are subjecting our bodies to enzyme-depleted foods. How, then, can we make up for this inevitable enzyme shortfall? Raw foods are a good approach. "Living" foods are better.

When a seed germinates, the enzymes that control the growth process are activated, and the chemical and biological complexity of the seed increases manyfold. Young cereal grasses are enzyme-rich, and they should be a mandatory part of any good diet.

Fresh vegetable juices provide instant access to all the enzymes of the cellular material of the plant. The cellulose in cell walls protects the cells from digestion until they have been broken down in our intestines. These cell walls have been shredded and separated in the juicing process.

As with many aspects of nutrition, there is danger in placing too much importance on only one aspect of food value. Note that many enzymes cannot do their work without vitamins and minerals. Sometimes these substances work catalytically as cofactors; sometimes they are involved in the reactions themselves. The inverse relationship is also true. Certain enzymes may be necessary for the utilization (and sometimes synthesis) of certain vitamins. So a good supply of vitamins and minerals — difficult to obtain in our industrialized world without resorting to supplementation — is essential. Vitamins can be synthesized. But remember that enzymes, at least for the time being, can be produced only by living organisms.

> The enzymes we can't synthesize in our own bodies have to come from food or from supplements derived directly from bioactive ingredients.

Personally, I'm not looking forward to the day we are able to manufacture enzymes artificially. Most likely, the result will be a big surge of interest in only a few isolated substances, while other enzymes and nutrients found in their natural context along with the few selected ones will be ignored. Nature always seems to be one step ahead of technology!

My intuition tells me that the Grass Juice Factor is a combination of many things, including enzymes, proteins, fats, and hormone precursors, and that biochemists won't have all the component parts identified for many years to come. Even when they do, I expect the expert advice to be the same as it is now: *Include young cereal grass in your diet because that ensures total nutrition!*

Anthony Cichoke, D.C., in his book *Enzymes and Enzyme Therapy*, offers suggestions for a "jump-start" enzyme dietary program. Here are a few of Dr. Cichoke's do's and don'ts (particularly the ones I advocate):

> ~ Eat fresh vegetables as much as possible.
> ~ Include a lot of garlic and onions in your diet.
> ~ Try sprouting.
> ~ Make sure ceramic dinnerware or crystal is lead-free.
> ~ Avoid common table salt, which may be an indirect enzyme inhibitor.
> ~ Include adequate bulk in your diet.
> ~ Drink large amounts of freshly made vegetable and juices.
> ~ Avoid excessively cold or hot beverages or foods.

Lifestyle change is not easy. We all know we're not supposed to eat pies, cakes, desserts, and candies, but most of us do anyway. Because of all the complexities involved in making healthful choices, the right kind of supplementation for enzyme health is vital.

Fermented Food Products

Through a different process, fermented foods add the digestive enzymes of the bacteria responsible for the fermentation. Yogurt (the health-store variety with live-culture bacteria), tofu, tempeh, and sauerkraut — again, the health-store kind that spares your digestive system a dose of sodium benzoate — are all superb foods in this category. They are easier to digest than their base ingredients; you might think of them as being partially digested already.

> Fermented food products contain enzymes that help with the assimilation and best utilization of many other foods and nutrients eaten during the same meal.

Food history reveals that societies that have enjoyed good health almost always include a fermented product in their daily diet. Bulgarians used yogurt; Germans, sauerkraut; Africans, fermented millet; Asians, fermented vegetables; and so on.

Because of the growing consensus of the beneficial attributes of fermented foods in *our* time and place, consuming friendly bacteria has become very popular. You may be aware of the Manchurian Mushroom craze that has swept the country. Also called kombucha, it is based on an ancient cultured tea. Those using this fungus report extraordinary health benefits. The reason? The fermented culture transforms the ingredients into enzymes!

"Viable" yogurt signifies the presence of live, friendly bacteria. Since heat can destroy the bacteria, yogurt should not be pastuerized following fermentations. In fact, federal law dictates that yogurt can only be so named if pasteurization precedes the fermentation process. Not all states follow this ruling, however. (Check with your local FDA office.) Yet another problem affecting live yogurt culture is the use of stabilizers and emulsifiers (tapioca and corn-starch are examples) in commercial yogurt products, which may inhibit access of the friendly bacteria in your intestines.

Supplementation of Lactobacillus acidophilus has been commonplace among the nutrition-aware in this country for many decades. Its values were promoted worldwide in 1908 by the Russian scientist Metchnikoff.

Acidophilus:

> ~ produces B vitamins
> ~ produces lactase enzymes and lactic acid
> ~ produces hydrogen peroxide
> ~ helps digest lactose
> ~ aids in food digestion in general
> ~ corrects digestive disorders
> ~ helps prevent bad breath
> ~ acts as a natural antibiotic
> ~ possesses cholesterol-lowering factors
> ~ inhibits or reduces yeast infections (candidiasis)

According to conclusions reached at an  international medical symposium in Sweden, friendly bacteria are of extreme importance in the nutrition of people all over the world. Fermentation is one of the most important functions of the colonic flora.

Chlorophyll

We breathe because the process of converting carbohydrates into energy consumes oxygen. All animal cells using energy require oxygen when they burn carbohydrates, giving off carbon dioxide as a product of this reaction. This process is called *respiration*. It's not unlike burning a log in the fireplace, in terms of what goes in and what comes out. Green plants, on the other hand, use *solar* energy. They actually *make* carbohydrates from nothing more than sunlight, water, and carbon dioxide — creating bio-organic substances from inorganic chemicals. That process is called *photosynthesis* and chlorophyll is the chemical that makes it possible. No wonder chlorophyll is known as "concentrated sunlight." Chlorophyll is also the pigment responsible for the deep green color of fast-growing plants.

An intriguing fact about chlorophyll is its chemical similarity to hemoglobin, the molecule that transports oxygen in your blood. Hemoglobin and chlorophyll share a structure called the *porphyrin ring*.

In hemoglobin, there's an iron atom bonded inside the ring; but chlorophyll has a magnesium atom in that very location instead.

It is tempting to speculate whether chlorophyll in food can be used to help manufacture new blood cells in our bodies. For more than a century, when the green pigment in plants was identified and called chlorophyll, scientists have puzzled over this question. It was early in this century that the similarity between hemoglobin and chlorophyll was confirmed. Although chlorophyll-rich plants contain many other nutrients that have been shown to be important for healthy blood, the *direct* use of the chlorophyll molecule has yet to be demonstrated.

Figure 4

HEMOGLOBIN

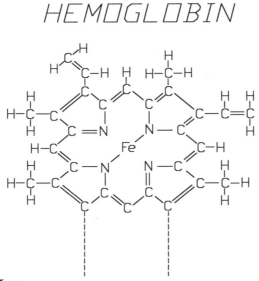

Fe = Iron

CHLOROPHYLL

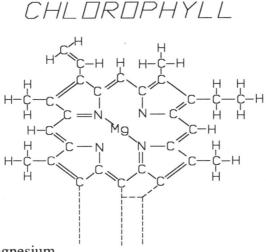

Mg = Magnesium

Note the similarity between our hemoglobin molecule and the plant's chlorophyll molecule.

Yoshide Hagiwara, M.D., in his book *Green Barley Essence, The Ideal Fast Food,* suggests a mechanism whereby chlorophyll may stimulate the production of hemoglobin by an indirect route. Although the intact chlorophyll molecule is difficult to absorb, there is a ring-opening process that facilitates this assimilation. The parts of the chlorophyll molecule are reassembled later as hemoglobin, as iron is substituted for magnesium. "Thus, the conclusion drawn from my reasoning," writes Dr. Hagiwara, "is that the green blood of the plant can become the red blood of man."[58]

For years chlorophyll additives have been used in many products. A form of chlorophyll, stabilized by reacting with copper and retaining a rich blue-green color, is totally useless biochemically. It is still used, however, as a synthetic food coloring, so don't be confused by processed food products containing chlorophyll in this form. It's not the same!

Algae is also a very potent and important source of chlorophyll.

In fact, more photosynthesis is performed by algae than by land plants. But few land animals eat algae as part of their natural diet. Most likely, this is due to the difficulties in breaking down the algae cell wall, hampering human cell absorption. But scientists have learned to break the cell wall in the laboratory! The finished product is sold in the marketplace as broken-cell or cracked-cell chlorella.

In his beautiful book, *Chlorella: Gem of the Orient*, Dr. Bernard Jensen writes:

> Chlorophyll is the most powerful cleansing and purifying agent in nature. It detoxifies the liver and bloodstream, and cleanses and sweetens the bowel.[59]

A derivative of chlorophyll has been used to treat a number of human conditions with no toxic effects reported. It was shown to have an inhibitory mechanism involving complex formation with carcinogens in the gut coupled with scavenging in the target organ.[60]

A study using the same derivative was conducted to show that chlorophyll contains an antimutagenic agent, proving to be effective against many carcinogens.[61]

Don't wait for the "human condition" — the disease state that wears you down, reduces life quality, and shortens lifespan. If chlorophyll is therapeutic, it is also preventive.

The best way to get chlorophyll in your diet is to eat organic dark green leafy vegetables, freshly picked.

Did I hear you groan? Second best is to eat carefully prepared special products made from these plants, designed to preserve the chlorophyll. (Convenient, easy, reliable, healthful: *kamut juice concentrate*.)

Chlorophyll, the perfect host for magnesium, is helpful in protecting against:

~ muscle cramping
~ menstrual problems
~ menopausal symptoms
~ cardiovascular disease
~ bronchial spasms
~ anemia
~ nonspecific, but widespread, therapeutic effects

Chlorophyll for Athletes

Chlorophyll has been called "The Green Steroid" because of its positive effect on muscle endurance.

> Green foods promote anabolic processes, the chemical reactions that build and repair tissue.

On the other hand, catabolic processes consume tissue for energy. In cases of extreme physical performance (when an athlete pushes to the limit, for example), it's the catabolic process that causes fatigue and tissue breakdown.

Remember a ship named the *Henrietta*? It was featured in Jules Verne's *Around the World in Eighty Days*. Running low on fuel in mid-ocean, the crew burned the entire superstructure in the boiler to keep on steaming ahead. The voyage was completed on schedule, but the ship looked more like an empty barge than a passenger cruise liner by the time it arrived at its destination.

> I think of the *Henrietta* whenever I see an athlete at the limit of exhaustion. He may be winning the race, but he's burning up his body to do it. A good supply of chlorophyll can delay that decisive moment — the moment of exhaustion.

8

WHEAT ALLERGIES

How Kamut Can Help

Agriculture Gone Astray
There is still uncertainty about how the hexaploid wheats
— the modern wheats, the group including common bread
wheat and most commercial wheats — appeared on the
scene some 8,000 years ago. In all likelihood, they did not
exist in nature prior to that time. They are the result of a
random cross-pollination between a diploid (with two sets
of seven chromosomes) and a tetraploid (with four sets of
seven). Two plus four equals six, so the result is the hexaploid group.

Random cross-pollination often occurs in nature, but it almost never produces a viable organism capable of reproducing itself. Even if the astronomically unfavorable odds
were overcome so that reproduction did occur, it's not likely
that the new hybrid would be competitive in the natural
environment. The new plant wouldn't have the benefit of
millennia of evolutionary selection — it would soon be
crowded out by the original species. That's the way nature
works.

Here's where human intervention could have entered the picture: 8,000 years ago today, a farmer could have been out tending two adjacent fields of grain. One field was planted with einkorn, the now forgotten diploid grain, and the other with tetraploid, possibly emmer or Persian wheat. The farmer noticed an unusual stalk with bigger kernels. She's seen these before — they appeared with regularity on the boundaries between the two fields — but she'd never had any success getting the seeds of these crossbreeds to germinate past the second generation.

This time, she collected the kernels and, on a hunch, planted them near kernels from other hybrids she'd collected.

"Maybe one combination isn't enough," she reasoned. "If I could get combinations of combinations, then I'd have more of different kinds of combinations, and perhaps eventually one of them would be fertile enough to plant a new field with these bigger kernels."

After several years of meticulous work (during which she was undoubtedly ridiculed by the skeptics in the clan), the farmer achieved some measure of success — a new line of wheat that could perpetuate itself, as long as people were around to plant the seeds, keep the einkorn and Persian varieties off the field, and scare the birds away.

Our hypothetical farmer, even though she was thinking only of feeding her family with the next harvest, had fired the starting gun for western civilization. She may also have launched her species on what some regard as a long downhill nutritional slide.

Why a downhill slide? Because modern wheat has been modified to such an extent that tens of millions of people in North America alone are allergic to it! We all know that bakery products are important ready-to-eat foods. But the nutritional quality of these products is low.

Among the current selection criteria for wheat are:

- ~ drought resistance
- ~ high yield
- ~ capacity to block pests and disease
- ~ suitability for mechanized harvesting
- ~ applicability for mechanized milling
- ~ ease of production and processing
- ~ color for pasta manufacturing

Nutritional content is far down the list — if it's on the list at all. The inferior nutrition is further accentuated by insecticide sprays, emulsifiers, leavening agents, bleaching agents, maturing agents, and finally, the high-tech applications used in producing the finished product.[62]

The result of this artificial selection is a biochemical concoction that has completely bypassed the natural selection process — something about which we can only guess. The further an agricultural product gets from its naturally evolved state, the more likely it is to cause allergic responses.

We don't fully understand the nature of food allergies and immune reactions. However, global studies, clinical observations, and food changes that occurred during war years allow us to speculate with a certain degree of accuracy. What we are absolutely sure of is that we have become dependent on a relatively small spectrum of wheat genes, and this has "done us in."

IMPORTANT!

The Making of a Slice of Bread

When flour is milled, it is yellow in color, making an unstable, inelastic dough, producing bread of low loaf volume. The flour must be oxidized for good texture — by allowing the flour to sit in air for several months or by the addition of additives. Potassium bromate and iodate may be added. To whiten flour, hydrogen peroxide, benzoyl peroxide, or lipoxidase are thrown in the pot to bleach the pigments and give a more desired color.

Milk solids help increase the elasticity of the dough. This also helps to form the crust flavor. Salt helps control bacteria so that only yeast will grow during fermentation. Sugar is added for flavor. Emulsifiers help to prevent staling. Calcium propionate controls moldiness.

There's more, much more, but I think you've gotten the point!

Nutritional composition of baked products can be improved by using quality wheat for milling. In the case of kamut versus common bread wheat, we do have hard data. One study, conducted over a six-month period, found that 70 percent of severely wheat-allergic participants had no significant reaction to kamut. Of the 100 food allergy patients tested, 70 percent showed greater sensitivity to common wheat than to kamut.

> "For most wheat-sensitive people," says Eileen Yoder, Ph.D., president of the International Food Allergy Association and head of this study, "kamut can be an excellent substitute for common wheat."

Anecdotal evidence is also beginning to accumulate. One health store owner (Kelly Goyen of Empirical Foods and Herbs, Cedar City, Utah) reports that, after selling kamut products to over 3,000 wheat-sensitive customers, there has not been a single report of allergic reactions. Although not a scientific study, it's definitely worthy of note.

Keep in mind that kamut does contain gluten, the protein component considered responsible for wheat sensitivities and malabsorption problems. Why is kamut more trouble-free?

Among the cereal flours, only wheat flour will form a dough suitable for breadmaking when mixed with water — the kind of breads and pastries to which Americans have become accustomed.

Three general properties of gluten proteins appear to be responsible for its ability to produce unique "light" bread products:

~ First, gluten's ability to form a cohesive dough.
~ Second, the ability of the dough to retain gas, unique to wheat.
~ Third, also unique to wheat, ability to "set" in the oven during baking, producing a rigid loaf.

The scientists fully understand the first two factors but know little about the third.[63] The dough's ability to become firm may be caused by a heat-induced cross-linking of the gluten proteins.[64] Therein may lie the rub: *Cross-linking of proteins has rarely been in our best health interest!*

For those who use the *vegetable* form of kamut, gluten sensitivity is obviously no problem. Nor is it a problem when consuming bakery products made from kamut. The reason for this benefit is not entirely clear. Is this because kamut has not been *hybridized*? We're not sure, but gluten toxicity (or lack of it) has been shown to be due to intrinsic differences in the grain.[65]

Of the 35 million people in the United States with food allergies, 10 percent are considered to have severe and immediate reactions, while the remaining 90 percent have delayed allergic reactions. That's the official line. Personally, I suspect that an even larger percentage of the population is adversely affected by subtle food allergies and that these allergies are responsible for chronic health problems, especially those that go unexplained for years. So even if you have no history of allergies, there's a good chance that eliminating foods known to cause a more immediate and more noticeable response in others will do a world of good for you.

Understanding Gluten Sensitivity

Gluten is the tough, viscid, nitrogenous substance remaining when the flour of wheat or other grain is washed to remove the starch. It's a gluey substance, also referred to as *wheat gum* — the insoluble protein constituent of wheat and other grains. *Gluten casein* is a protein resembling casein, present in gluten.

 The similarity of the protein components of milk and wheat may be the reason why those who are sensitive to one are often sensitive to the other.

Like so many other health conditions, it's not just an *either or* problem. Some people may be severely sensitive to gluten; others just slightly reactive. To add to the confusion, sensitivities are also related to dose, frequency, and age. This explains why you may be sensitive one week but not another, or why you think you suddenly developed a sensitivity although you never had one before. To further complicate matters, reactions manifest themselves in a myriad of ways — from fatigue to headache to diarrhea to bloat — with the cause not always apparent.

To date there is no definition as to what amount of gluten in the diet might be tolerable. I suspect it's a variable. For those who do have severe problems, holding to a totally gluten-free diet can be difficult, particularly for adolescents. How do you convince a teenager to say *no* at the pizza parlor? It's not even easy for adults to toe the nutritional straight and narrow! A diet excluding common commercial wheat, oats, rye, and barley, however, is the cornerstone of the treatment for the wheat-sensitive.[66]

If you do attempt such a diet and find it doesn't work, chances are its because of hidden gluten intake. Gluten is found in the additives and fillers used in many medications, in cocktails, and in beer and whiskey. Most brands of ice cream contain gluten, as do many dairy products, chewing gum, and even communion wafers.[67] We eat mystery meals, don't we — we don't know what we're eating!

Studies done with test animals show an underlying *permeability* abnormality involved in the gluten sensitive.[68] Other research shows that bone derangements are common.[69] Measurement of gluten antibodies is recommended in the diagnostic evaluation of children with short stature,[70] mainly because slow growth rate may be linked to the malabsorption incurred by the gluten in wheat.[71]

Gluten intolerance is difficult to overcome. We do *not* outgrow it, as many mistakenly think. Symptoms manifest themselves differently as we age. Reintroduction of gluten produced recurrence of gluten intolerance in 96 percent of afflicted children treated with a gluten-free diet for two to three years. Recurrence of the problem was accompanied by the atrophy of the intestinal villi.[72]

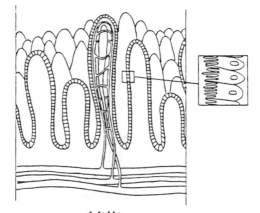

Villi
The absorptive surface of the intestinal tract.

Your digestive system is a large ecological field. Villi, the tiny hairlike structures that reach out to absorb nutrients in your small intestine, can, if stretched, cover an entire football field.

Villi flatten out when you're not good to them — when you subject them to foods you cannot properly absorb.

In the gluten-sensitive, when gluten comes in contact with the lining of the intestine, the lining blocks the absorption of gluten plus virtually every other nutrient eaten with it. So your ability to utilize nutrients from other foods is also diminished! In addition, you suffer a reduced protective capacity against oxidizing stresses.[73]

Children fed cow's milk protein between the second and third months of age frequently have gluten intolerance. A close correlation between feeding and the development of food intolerance has been noted in many studies.[74]

Gluten in the maternal diet primes an immune response in the offspring, as demonstrated in animal testing. These findings indicate that sensitization to maternal dietary antigens readily occurs in utero or shortly after birth.[75]

We're back to square one: If you even suspect that you are wheat-intolerant, consider cereal *grass* rather than cereal *grain*. Kamut juice concentrate should help to cajole your healing forces into action. If you must have baked goodies, buy or bake with kamut as your grain of choice.

KAMUT: Why It Is a Super Grain

Organically grown kamut grain and flour are marketed through companies in the United States, Canada, and Europe. You should be able to find whole kamut grains sold by the pound in the bulk food section of your local health store. Many distributors and manufacturers use kamut grain and flour to produce kamut products such as pasta, sprouted bread, and kamut cereal. There's even a children's breakfast cereal called *Kamutios*!

Most of these products use kamut in place of our worn-out wheat strains.

> If you're using kamut as a vegetable food (the way I believe it is most valuable), you should look for the dehydrated leaf material or juice concentrate from the leaves of the young kamut plant — the result of the growing process described earlier.

If you are not ready to make the drastic dietary and lifestyle changes necessary to accommodate a diet based on green vegetables (and who really is?), a prepared supplement like dehydrated kamut juice could be the best solution. It's a remarkably efficient way to get high-potency vitamins and minerals along with the not-so-well-isolated factors discussed earlier. Even though you'll be dissolving powder in liquid, think of these products as concentrated food rather than vitamins. Nutrient-rich, closer-to-natural supplements (as compared with synthesized) remain, far and away, the best way to add vitamins, minerals, and enzymes to your sadly depleted foods. Tending the wounds of the injured does nothing to reduce the number of casualties. *You must correct the causes!*

Even though most people don't have access to organic produce or the time to juice every day, home juicer sales have exceeded a record one billion dollars. Ever since we have been including kamut's concentrated leafy green juice as a regular part of our supplement regimen, dust has gathered on the two juicers sitting in our pantry.

This product is at the peak of my *nutritionally powerful but convenient* list.

The only two high-energy ingredients in kamut concentrate are the juices from young, tender leaves of kamut and alfalfa plants — 65 percent kamut and 35 percent alfalfa. Like other top-of-the-line nutritional products, there are no fillers, binders, starches, or dextrin.

The kamut and alfalfa are planted together so that the alfalfa builds nitrogen in the soil, which in turn makes for higher chlorophyll content in the kamut. Explained earlier (but it bears repeating), the unique on-site processing harvests, rinses, juices, and dries the leaves as quickly as possible. And the drying occurs at only 88° Fahrenheit! (Enzymes may be rendered inactive at 118°, and some even at 105°.) Annual crop rotations allow each field to be harvested only once a year, thereby providing maximum nutrition year after year. Recall, too, that it is grown on an ancient volcanic lake bed, watered with the subterranean and mountain spring water, totally organic and never subjected to pollutants or chemicals in its entire history.

Alfalfa, known for centuries as the father of all food, has long been among my favorites because its roots, rotor-rooter style, crawl deep into the earth to capture vital nutrients from the soil.

Alfalfa contains a high amount of chlorophyll, beta-carotene, and all the other wonderful nutrients that spill out from soil and sun to plant — to help us experience true well-being.

And there's more: kamut is wrapped in 100 percent recyclable packaging! *Now* are you ready to join that Kamut Lovers Club?

Various manufacturers are taking advantage of the dehydrated kamut-alfalfa formula by creating mix-and-match new products. One company is combining the powdered juice with other cereal grasses and wonderful sprouts. The ingredients are dehydrated green kamut juice, green barley juice, alfalfa juice, wheatgrass juice, oat grass juice, lecithin, cracked-cell chlorella, broccoli, spinach, parsley, dandelion greens, kale, cabbage, apple pectin fiber, acidophilus, sea dulce, sea vegetables, and a few enzymes.

Wow! They call this *Pure Green Power,* and that's exactly what it is!

Another company adds yucca root, flax seed, brown rice, and spirulina to the medley. Yet another manufacturer combines dried kamut juice with garlic. And another is mixing this already-marvelous product with acidophilus! (The best gets even better!)

We've come a long way since I lined up the synthesized single-nutrient vitamin preparations on my kitchen counter!

Green Kamut

9

BEING YOUR OWN DOCTOR

Helping Yourself to Health

Everyone's Unique
The great Roger Williams taught us about individual bio-chemistry. He said:

> Each member of the human family is a unique specimen with respect to his or her pattern of metabolic activities and the quantitative aspects of his or her nutritional requirements. These differences are often far from trifling and the only kind of people who need or can use nutrition are *individual* people.

Note these few examples of individual differences:

~ The total weights of livers vary about fourfold.
~ As blood loops through the aorta, it may go through one, four, five or six branches.
~ Pain spots, which are highly sensitive, vary on each person's hand.
~ Hearing in "normal" ears is reflected in sensitivity to a variety of different pitches.

The point is that no book about food should ever give medical advice, no matter who writes it. Yet the reality is that
every day millions of people read books about health and
nutrition and make personal health choices without consulting a physician. In an ideal world, each of us would
have a good working relationship with a physician who
shares our own philosophy and who would give us the kind
of professional guidance we require for optimal health.
Obviously, we are far from this ideal. In one widely quoted
survey, published in the *New England Journal of Medicine*,
it was shown that a large percentage of Americans sought
alternative health therapies (including nutritional regimens)
and *didn't tell their doctors!* This isn't necessarily negative, since most physicians are woefully undereducated
about nutrition. When you attempt to heal yourself with
food — even if you have a good health plan and a compe-
tent physician — chances are you will have only limited
support, if any, from the medical establishment.

What do you do if your doctor is not supportive of alternative nutrition-based approaches to health? Or, like an alarm-
ingly high percentage of the population, what if you have
no regular doctor at all and no practical means of consulting with a physician, short of a life-threatening emergency?

Many of us have no choice but to be our own doctors, which
is not as bad as it sounds at first. Keep in mind that, even
if you had the best doctor in the country, the ultimate responsibility for your health is yours. You put the food in
your mouth, you decide when (or whether) to exercise, and
you can limit (or not) many of the toxins you ingest. There
is no magic to make up for a lifetime of bad habits. You
don't poison yourself to health.

I Have a Salad Every Day. Isn't that Enough?

No! Most salads, at least most salads eaten in this country, are made with iceberg lettuce, also known as head lettuce. Head lettuce is just about the most worthless vegetable on the table, and compared with other lettuce varieties, it is a junk food. "What?" I hear you say. "Lettuce, a junk food? Explain this!"

For starters, lettuce is mostly water. Okay, no big deal. But it's also very low in the all-important chlorophyll because the head construction prevents sunlight from being harnessed and reaching most of the leaves. Makes sense — there's no reason for the plant to have chlorophyll in its leaves if there's little light for photosynthesis.

But head lettuce has been genetically engineered almost beyond cellular recognition. Grown with large amounts of pesticides and sprays to preserve crispness, it's groomed so it won't look "tired" after its journey to restaurants and supermarkets across the country. With nutrients absent and toxins present, if iceberg lettuce appears fresh, it's only an illusion.

> America, wake up! Iceberg lettuce in your salad is not only poor nutrition but out of style.

Iceberg lettuce should go the way of the marshmallows that used to be so popular in fruit salad. Let the food industry work a little harder in our behalf.

Sure, you could get organic head lettuce. But why not use butterhead (Boston and Bibb), Cos (romaine), leaf lettuce, or even spinach, kale, or any one of a dozen tastier and more nutritious alternatives?

Having chosen a healthful salad variety, you also have to be careful with the salad dressing. Watch out for products containing polyunsaturated vegetable oils. Oil in salad dressing is almost guaranteed to be at least partially oxidized or rancid as a result of processing, storage, and exposure to heat, light, and air. What's more, your body might take advantage of the quick energy fix from the oil, carbohydrate, or sugar in the salad dressing and fail to fully utilize the green stuff. (We have found one delightful exception, *grape seed oil*, which is the most stable of all oils available.)

Will you have a salad left after you banish the head lettuce and the dressing? If you do, you might have some good stuff in your salad bowl. Fresh carrots, for example. But are they thin-sliced or shredded? And how long have they been like that, exposed to the air? This goes for the cucumbers and green peppers. The very process of preparing a salad takes time, leaving the vegetables unprotected and exposed to air and light in potentially deleterious ways.

The more thinly the veggies are sliced, the more likely oxidation erodes their vitamins and enzymes. (Minerals are pretty safe here.) By the time they get from the big farm to the supermarket to your grocery bag to your refrigerator to your table, they're already considerably degraded. Go heavy on sprouts and leafy greens in that salad because they don't have to be mushed, mashed, or mangled before you serve them.

Fortunately, most vegetables have a wonderfully complex and "smart" skin structure that keeps air and pathogens from "breaking and entering."

In my own clinical experience, one piece of dietary advice seems to have the most impressive results: *Cut out all foods containing wheat and dairy products.* Two weeks on this regimen appears to make almost anyone feel better! Admittedly this is very hard to do in this country. Commercial wheat is everywhere. But if you *really* want to improve your health (many people don't, when it comes down to it), then you *will* make the effort. Travelers to the Far East, where both wheat and milk are normally absent from the local cuisine, often report an unexpected boost of energy!

A more extreme strategy *always* works for me and for anyone I can persuade to follow it: *Eat every meal with high amounts of complex carbohydrates and a small amount of quality protein.* That translates to lots of vegetables and, if you are not a vegan, very small amounts of fish or fertile eggs. This is my personal therapeutic/recovery diet, guaranteed to put me back in top condition after any kind of stress. As I return to my "normal" diet, I attempt to keep track of what makes me feel strong and healthy, what makes me feel lethargic and weak. It's a slow and imperfect process, but after awhile the puzzle pieces begin to fit together.

The consummate program for someone who is ill involves detoxification with many glasses of kamut juice concentrate (stirred into pure water) throughout the day, as much vitamin C as you can tolerate, some acidophilus, *and nothing else.*

Several days on this regimen may help rid your body of abscesses, cysts, and who knows what other unwanted foreign invaders! I know this works from much personal and clinical experience! For ongoing or more serious problems, please see your physician.

Kamut grain shows promise as a form of wheat that can be introduced without triggering allergic responses that cause so much widespread suffering in such a large portion of the population.

Of course, there are many other important nutritional principles to keep in mind. I could write a book about it. (And in fact I did! See the list of my books.) My advice — culled from a half century of clinical observation, from interviews with the world's noted experts in the field, and from extensive research, can be summarized very briefly: Eat as much of your food as possible in the form closest to how it was when it was growing, flying, running, or swimming. That is, minimize processing and cooking! Eat *lots* of vegetables. As long as leafy greens are a staple in your diet, odds are you'll be well supplied with whatever vitamins, minerals, or trace nutrients are in fashion this year, next year, or any year, without even thinking about it. And if you can't get the real thing, second best is okay, too — use supplements that contain extractions of young green shoots.

Avoid junk. Junk is primarily sugar (you knew that), and junk includes fats and oils that have been cooked at high temperature (fried) or allowed to become even slightly rancid (vegetable oil from the supermarket — maybe you didn't know that). Junk is food processed with oil. Junk is most cooked meat. Junk is any alcoholic, caffeinated, or carbonated beverage. Junk is almost all packaged snack food. If it's crispy and it didn't grow crispy, it's junk.

Did I just ruin your day? If I did, I may also have just saved your life!

As high technology continues to offer foods prepared with higher temperatures and lower nutrition — impairing the very structure of the food — there is evidence of only very modest improvement in the American diet during the past twenty years. According to the National Center for Chronic Disease Prevention and Health Promotion, changes must occur at a faster pace if the *Year 2000 Dietary Goal* (which advises that we consume five or more daily servings of fruits and vegetables) is to be met.[76] Researchers at the medical college of Augusta, Georgia, suggest making fruits and vegetables more available.[77] No easy task!

Without resources dedicated to dietary modification in the general population, it is unlikely that the potential savings of more than 300,000 new cancer cases, 160,000 deaths, and the $25 billion in associated costs will be realized in the foreseeable future. At least thirty-five percent of cancer cases (if not more) are associated with one lifestyle practice: SAD, the Standard American Diet.[78]

Nobody can be expected to turn the clock back 8,000 years and eat only fresh-picked, fresh-killed, and fresh-sprouted natural foods. That's where special concentrated products come in. Kamut grass is particularly suited to this form of supplementation. You can enjoy the benefits of this remarkable food without a major assault on your lifestyle. If a magic bullet did exist, it would be kamut juice concentrate.

Consider the positive. There are lots of great-tasting, natural, healthful, unmolested, wholesome foods that are just waiting to be discovered and enjoyed. While you're prospecting through the cookbooks and health stores, protect yourself with easy-to-take, convenient, concentrated green power for a steady renewal of force.

Endnotes

Kamut juice concentrate is for real people in a real world. It is for:

~ Pregnant women — because of kamut juice concentrate's smorgasbord of incredible nutrients. There is no other time when good nutrition is as important.

~ Lactating women — because the quality of human milk can dictate the physical and mental quality of the baby who is fed this bounty.

~ Vegetarians — because of kamut juice concentrate's unusual amino acid content, lacking in almost all vegetable foods.

~ Children — because they don't like vegetables.

~ Athletes (the serious kind) — because of kamut juice concentrate's natural steroid and anabolic advantages.

~ Athletes (the Sunday-only variety) — because they are more subject to muscle cramps, which may be relieved on a steady diet of kamut juice concentrate.

~ Those whose stomachs are ill-adapted to the turn their bad habits have taken — because kamut juice concentrate is so easy to digest.

~ Senior citizens — because they often have digestive problems and don't bother to cook, especially if they live alone; and because metabolic efficiency slows down as we age, so we need all the help we can get. (Just a reminder: tomorrow you will be one day older than you are today!)

~ Women concerned about PMS, menopausal symptoms, and osteoporosis — because kamut juice concentrate helps to supply hormonal nutrient precursors.

~ Anyone recuperating from any illness — because kamut juice concentrate speeds the healing process.

~ Anyone on weight-loss programs — because, on most diets, we lose nutrients along with the pounds, and kamut juice concentrate can help to replace them (the nutrients, not the pounds).

~ The coffee-impaired or the meat eaters, whose acid-alkaline ratios are askew — because kamut juice concentrate offers the pH balance we all require for health.

~ Those who dine on coronary catastrophes, because of kamut juice concentrate's antioxidant protection — helping to cast off our layers of the synthetic and the artificial.

~ Those who do not want to get tired or old before their time, for reasons already stated.

~ And finally, those who just want to be fortified against the anxieties of the coming day — also for reasons already stated.

Anyone not mentioned above, stand up.

The sun is shining.
It is up to you to move into it.
Illness is no fun.
Please take care of yourself!

REFERENCES

1 Brownowski, Jacob, 1973. *The Ascent of Man*, Little, Brown, & Co., Boston & Toronto.

2 Dessi M; De Vincenzi M; Maialetti F; Mancini E. "Effect of alpha-gliadin-derived peptides from bread and durum wheat on K562(S) cells." *Italian J of Gastroenterology*, 1992 Sep, 24(7):397-9.)

3 Montalbano, W.D. *New York Times*, December 23, 1993.

4 Peleg M. "Darwinian evolution patterns in food products and beverages."Dpt of Food Science, Univ of Mass, Amherst 01003.
Crit Rev Food Sci Nutr 1994;34(2):95-108.

5 Seibold, RL. *Cereal Grass, What's In It For You!* (Lawrence, Kansas: Wilderness Community Education Foundation, 1990), p 13.

6 Ibid, p 15.

7 Willett WC. "Diet and health: what should we eat?" *Science*, 1994 Apr 22;264(5158):532-7.

8 Bal DG, Foerster SB. *Cancer*, 1993 Aug 1;72(3 Suppl):1005-10.

9 Taylor A. "Cataract: relationship between nutrition and oxidation." *J Am Coll Nutr*, 1993 Apr;12(2):138-46.

10 "Effects of fruit juices, processed vegetable juice, orange peel and green tea on endogenous formation of N-nitrosoproline in subjects from a high-risk area for gastric cancer in Moping County, China." *Eur J Cancer Prev*, 1993 Jul;2(4):327-35.

11 Cournot MP, Hercberg S. "Prevention of mineral deficiencies (iron, calcium and magnesium)." CNAM, Paris. Rev Prat, 1993 Jan 15;43(2):141-5.

12 Mangels AR et al. "The bioavailability to humans of ascorbic acid from oranges, orange juice and cooked broccoli is similar to that of synthetic ascorbic acid." *J Nutr*, 1993 Jun;123(6):1054-61.

13 Bulux J et al. "Plasma response of children to short-term chronic beta-carotene supplementation." *Am J Clin Nutr*, 1994 Jun;59(6):1369-75.

14 Feldman EB. "Dietary intervention and chemoprevention—1992 perspective." *Prev Med*, 1993 Sep;22(5):661-6.

15 Skowronski RJ; Peehl DM; Feldman D. "Vitamin D and prostate cancer: 1,25 dihydroxyvitamin D3 receptors and actions in human prostate cancer cell lines." *Endocrinology*, 1993 May, 132(5):1952-60.

16 Feher J; Pronai L. "The role of free radical scavengers in gastrointestinal diseases." *Orvosi Hetilap*, 1993 Mar 28, 134(13):693-6.

17 Gey KF; Stahelin HB; Eichholzer M. "Poor plasma status of carotene and vitamin C is associated with higher mortality from ischemic heart disease and stroke: Basel Prospective Study." *Clin Inv*, 1993 Jan, 71(1):3-6.

18 Roncucci L et al. "Antioxidant vitamins or lactulose for the prevention of the recurrence of colorectal adenomas." *Diseases of the Colon and Rectum*, 1993 Mar, 36(3):227-34.

19 Rohan TE et al. "Dietary fiber, vitamins A, C, and E, and risk of breast cancer: a cohort study." *Cancer Causes and Control*, 1993 Jan, 4(1):29-37.

20 Zheng W et al. "Serum micronutrients and the subsequent risk of oral and pharyngeal cancer." *Cancer Res*, 1993 Feb 15, 53(4):795-8.

21 Steinmetz KA et al. "Vegetables, fruit, and colon cancer in the Iowa Women's Health Study." *Am J Epidemiol*, 1994 Jan 1;139(1):1-15

22 Goodman GE. "Chemoprophylaxis strategies in high-risk groups with an emphasis on lung cancer." *Chest*, 1993 Jan, 103(1 Suppl):60S-62S.

23 Lippman SM; Benner SE; Hong WK. "Chemoprevention strategies in lung carcinogenesis." *Chest*, 1993 Jan, 103(1 Suppl):15S-19S.

24 Herzog F, Farah Z, Amado R. "Nutritive value of four wild leafy vegetables in C'ote d'Ivoire." *Int J Vitam Nutr Res*, 1993;63(3):234-8.

25 Rahman MM et al. "Can infants and young children eat enough green leafy vegetables from a single traditional meal to meet their daily vitamin A requirements?" *Eur J of Clin Nut*, Jan 1993;47(1):68-72.

26 Bottcher H. "The assessment of nutrient value of vegetables." *Nahrung*, 1993;37(1):20-7.

27 Rauma AL et al. "Effect of a strict vegan diet on energy and nutrient intakes by Finnish rheumatoid patients." *Eur J Clin Nutr*, 1993 Oct;47(10):747-9

28 Rizzo AF et al. "Protective effect of antioxidants against free radical-mediated lipid peroxidation induced by DON or T-2 toxin." National Veterinary and Food Research Institute, *Zentralbl Veterinarmed*, [A] 1994 Mar;41(2):81-90.

29 *N E J of Med*, Feb 2, 1995.

30 Lee HP. "Diet and cancer: a short review." *Ann Acad Med Singapore*, 1993 May;22(3):355-9

31 Mayer JP et al. "Consumption of fruits and vegetables in Missouri." *Missouri Med J*, 1993 Oct;90(10):653-5

32 Kant AK, Schatzkin A. "Consumption of energy-dense, nutrient-poor foods by the US population: effect on nutrient profiles." *J Am Coll Nutr*, Jun 1994, 13(3):285-91.

33 Wolfe WS, Campbell CC. "Food pattern, diet quality, and related characteristics of schoolchildren in New York State." *J Am Diet Assoc*, 1993 Nov;93(11):1280-4.

34 McMahon J, Parnell WR, Spears GF. "Diet and dental caries in preschool children." *Eur J Clin Nutr*, 1993 Nov;47(11):794-802.

35 Lopez I et al. "Breakfast omission and cognitive performance of normal, wasted and stunted schoolchildren." *Eur J Clin Nutr*, 1993 Aug;47(8):533-42.

36 Thompson FE, Dennison BA. "Dietary sources of fats and cholesterol in US children aged 2 through 5 years." *Am J Public Health*, 1994 May;84(5):799-806.

37 Martinez Belchi A et al. "Approximation to the nutritional status with a nutrition survey in the health area." *Aten Primaria*, 1993 Feb 1;11(2):81-3

38 Nestle M. "Food lobbies, the food pyramid, and U.S. nutrition policy." *Int J Health Serv*, 1993;23(3):483-96.

39 Whorton JC. Department of Medical History and Ethics, University of Washington, Seattle 98195. *Am J Clin Nutr*, 1994 May;59(5 Suppl):1103S-1109S.

40 Haas, EM, *Staying Healthy With Nutrition* (Berkeley, CA: Celestial Arts, 1992), p 728.

41 Levine S, Kidd PM, *Antioxidation Adaptation, Its Role in Free Radical Pathology*, (San Leandro, CA: Allergy Research Group, 1986), p 294

42 Rowell M. "Eradication of vitamin A deficiency with 5 cents and a vegetable garden." *J Ophthalmic Nurs Technol*, 1993 Sep-Oct;12(5):217-24

43 Quillin P. *Healing Nutrients* (New York: M. Evans & Co., Inc., 1990), p. 264.

44 Guthrie H. *Introductory Nutrition* (St. Louis, MI: C.V. Mosby Company, 1983).

45 *J of the Natl Cancer Inst*, November 1, 1989, 81:21.

46 Rosenfeld, I. *Doctor, What Should I Eat?* (New York: Random House, 1995), p 50.

47 *Intnl J of Epid*, Dec 1990.

48 *Preventive Medicine*, Jan 1993.

49 *Cancer*, Dec 1992.

50 *Anticancer Res*, Sep 1990: "

51 *Voprosy Onkologii*, 1992, 38:141.

52 Jensen, B. *Seeds and Sprouts for Life* (Escondido, CA: Hidden Valley Health).

53 Moore R, Webb G. *The K Factor* (NY: Macmillan, 1986), p 215.

54 Scala J. *High Blood Pressure Relief Diet* (NY: Penguin, 1990) p 52.

55 Rhoades R, Pflanzer R, *Human Physiology* (NY: Harcourt Brace Jovanovich, 1992) p 218.

56 Cichoke, A. *Living Naturally, Let's Live,* 1994.

57 Seibold, RL. *Cereal Grass, What's In It For You!* (Lawrence, Kansas: Wilderness Community Education Foundation, 1990), pp 24-25.

58 Hagiwara Y, *Green Barley Essence, The Ideal Fast Food* (New Canaan, CT: Keats Publishing, 1985), p 69-70.

59 Jensen B. *Chlorella: Gem of the Orient* (Escondido, CA: B Jensen, Publisher, 1987).

60 Dashwood RH; Breinholt V; Bailey GS. "Chemopreventive properties of chlorophyllin....*Carcinogenesis,* 1991 May, 12(5):939-42.

61 Wu ZL et al. "Antitransforming activity of chlorophyllin against selected carcinogens and complex mixtures." *Teratog Carcinog Mutagen,* 1994;14(2):75-81.

62 Chavan JK, Kadam SS. "Nutritional enrichment of bakery products by supplementation with nonwheat flours." *Crit Rev Food Sci Nutr,* 1993;33(3):189-226

63 Hoseney RC. "Wheat gluten: rheological and gas retaining properties." *Adv Exp Med Biol,* 1991;302():657-66

64 Hoseney RC, Rogers DE. "The formation and properties of wheat flour doughs." *Crit Rev Food Sci Nutr,* 1990;29(2):73-93.

65 Dessi M et al. "Effect of alpha-gliadin-derived peptides from bread and durum wheat on K562(S) cells." *Ital J of Gastroenterology,* 1992 Sep, 24(7):397-9.

66 Auricchio S, Troncone R. "Effects of small amounts of gluten in the diet of coeliac patients." *Panminerva Med,* Apr-Jun 1991, 33(2):83-5.

67 Rosenfeld, I. *Doctor, What Should I Eat?* (New York: Random House, 1995), p 308.

68 Hall EJ, Batt RM. "Abnormal permeability precedes the development of a gluten sensitive enteropathy in Irish setter dogs." *Gut,* Jul 1991, 32(7):749-53.

69 Molteni N et al. "Bone mineral density in adult celiac patients and the effect of gluten-free diet from childhood." *Am J Gastroenterol,* Jan 1990, 85(1):51-3

70 Knudtzon J, Fluge G, Aksnes L. "Routine measurements of gluten antibodies in children of short stature." *J Pediatr Gastroenterol Nutrition,* Feb 1991, 12(2):190-4.

7 Hernandez M et al. "Growth in malnutrition related to gastrointestinal diseases: coeliac disease." *Horm Res,* 1992;38 Suppl 1():79-84

72 Kaczmarski M. "The provocation test in children with cow-milk protein and gluten intolerance: evaluation of the clinical response and lesions in the mucous membrane of the small intestine." *Pol Tyg Lek,* Feb 1990, 45(8-9):161-5.

73 Boda M, Nemeth I. "Decrease in the antioxidant capacity of red blood cells in children with celiac disease." *Acta Paediatr Hung,* 1992;32(3):241-55

74 Kaczmarski M, Kurzatkowska B. "The contribution of some environmental factors to the development of cow's milk and gluten intolerance in children." *Rocz Akad Med Bialymst,* 1988-89;33-34():151-65

75 Troncone R, Ferguson A. "In mice, gluten in maternal diet primes systemic immune responses to gliadin in offspring." *Immunology,* 1988 Jul;64(3):533-7

76 Byers T. "Dietary trends in the United States. Relevance to cancer prevention." *Cancer,* 1993 Aug 1;72(3 Suppl):1015-8

77 Domel SB et al. "Measuring fruit and vegetable preferences among 4th- and 5th-grade students." *Prev Med,* 1993 Nov;22(6):866-79.

78 Bal DG, Foerster SB. "Dietary strategies for cancer prevention [see comments." *Cancer,* 1993 Aug 1;72(3 Suppl):1005-10

Index